PROFITING FROM INDUSTRY 4.0

THE ROAD TO FUTURE VALUE IN MANUFACTURING

Profiting from Industry 4.0: The road to future value in manufacturing

Published by Novaro Publishing Ltd, Techno Park, Coventry University Technology Park, Puma Way, Coventry CV1 2TT e: publish@novaropublishing.com.

ISBN: 978-1-8380674-6-5
E-ISBN: 978-1-8380674-7-2

A CIP record for this book is available from the British Library.

Designed by Chantel Barnett, Clear Design CC Ltd

For further details about our authors and our titles, see:
www.novaropublishing.com

PROFITING FROM INDUSTRY 4.0

THE ROAD TO FUTURE VALUE IN MANUFACTURING

MARCOS KAUFFMAN

CONTENTS

Part 3: Profiting from Industry 4.0

FOREWORD

The internet of things (IoT) and the industrial internet of things (IIoT), also known as Industry 4.0, are part of the expansive digital technologies mix that is transforming every economic sector. IoT, essentially, refers to a wireless network of connected devices, objects and sensors that collect and communicate information. Assets and machines can be turned into IoT devices if they have the capability of being connected to the internet, either to communicate data or control it.

Today, innovations in industrial software are being powered by artificial intelligence-based analytics that are accessible via networks of fifth generation communication technologies (5G), interacting with a range of cloud and edge-computing services. These converging technological advancements are providing valuable and adaptive packaged business capabilities for manufacturers, helping them unlock drastic reductions in downtime, secure better operational efficiencies and forge more customer-centric experiences.

Industry 4.0 has multitudes of applications in manufacturing. It can facilitate the production flow in a manufacturing plant, as IoT devices automatically monitor development cycles and manage warehouses, as well as inventories. Operated across an enterprise's ecosystem, Industry 4.0 also has the potential to transform the enterprise's supply chains by offering, amongst other benefits, the following:

- Better visibility and tracking in real time
- End-to-end transparency and accountability
- Accelerated product lifecycle management
- Intelligent production planning and detailed scheduling
- Focused process automation and robotics utilization
- Improved dynamic warehouse management
- More accurate demand forecasts and inventory positions

For example, a food manufacturer could use Industry 4.0 to determine if their products have been exposed to temperatures, pressures or other environmental conditions that may render the food unsafe for consumption, and hence, avoiding the chance of product recalls and the consequential hit on the profit line.

Embracing Industry 4.0 in a manufacturing business requires a shift in both the business and the operating models of an organization. A leadership intervention inspired by actionable commitment will ensure that the strategic approach to Industry 4.0's deployment is both considered and accurately implemented.

Data, or more specifically big data, generated and captured across the value chains remains the key foundational element that enables smart and automated capability. Over this decade the interplay between artificial intelligence and Industry 4.0 will continue to grow, yielding better solutions in areas of security, stakeholder interactions and predictions. The availability of cost-effective sensors such as Lidar (low-cost solid-state light detection and ranging), as well as plastic, biological and chemical ones, together with sensor analytics, wireless networks and innovative low-power IoT chips will further disrupt manufacturers, resulting in the emergence of highly differentiated business models that generate multi-modality customer experiences.

Whilst the opportunities presented by the confluences of Industry 4.0 are undeniably compelling, the rate of take-up by Industry 4.0 by manufacturers remains patchy or, in many cases, unscalable, thus failing to demonstrate how business value can be derived.

This book, Profiting from Industry 4.0: The road to future value in manufacturing helps manufacturers to find new ways to create value based on a capacity to change and learn. It offers a set of tools, templates and guidelines to improve strategic and operational decision-making processes, to build a robust business case for investing in, or leveraging, Industry 4.0 technologies. It gives guidance on how to establish solution paths with confidence and how to utilize Industry 4.0 technologies to drive and deliver sustainable growth.

PROFESSOR SA'AD SAM MEDHAT

Chief executive, Institute of Innovation & Knowledge Exchange, and visiting professor of innovation and digital transformation, University of Westminster, PhD, MPhil, CEng, FIET, FCIM, FCMI, FRSA, FIKE, FIoD, FIRL

INTRODUCTION

Digital technologies dominate today's society in a manner that we could not have anticipated when the first digital computers emerged in the 1950s. Nowadays, we take for granted that personal computers, tablets, mobile phones and other digital devices are ubiquitous in our day-to-day lives. We do not think twice about streaming our favourite books, music and films, participating in online communities in which we create, share and consume content or through which we buy, sell and exchange goods at the click of a button. We also live in a world where we collectively generate huge volumes of data, which can be mined for patterns and from which algorithms can predict behaviour or generate new content. In short, we are in the midst of a digital revolution – and have been for at least a couple of decades. It brings with it many opportunities and challenges to which the industrial sectors are not immune.

This was pointed out by Klaus Schwab, the founder and executive chair of the World Economic Forum, who in January 2016 declared that the world was entering a fourth industrial revolution[1]:

> We stand on the brink of a technological revolution that will fundamentally alter the way we live, work and relate to one another. In its scale, scope, and complexity, the transformation will be unlike anything humankind has experienced before.

This book focuses particularly on manufacturing industry which is in the midst of a digital revolution, referred to as Industry 4.0, where factories and whole value chains are becoming digitalized, connected and visible. As a result, innovation in manufacturing is speeding up and becoming more open. Data is being shared and value is being created in real time. Businesses' performance can be transformed and new markets created, either by existing players or by new disruptive ventures.

For all the excitement of Industry 4.0, the risks are too often overlooked. Like other digital markets, value can easily just flow to the top of the chain, leading everyone else to struggle as commodities. In this scenario, businesses down the value chain can invest in new technologies, but still find themselves struggling to profit from innovation, as many manufacturers now are. The challenge is to understand how value is now created, capture its dynamics, begin to play by the new rules and secure a healthy return on investment.

This book gives those on the manufacturing frontline a holistic view, a set of tools and a series of guidelines with which they can start to capture value from their investment in Industry 4.0 technologies and take a commercial lead in their value chains.

Based on a comprehensive review of how manufacturing business models and contracts currently operate, it highlights a number of questions for manufacturers to ask and reviews their options for managing innovation, designing business models, managing intellectual property (IP) and gaining a lasting source of competitive advantage.

This book is organized in three main parts. In Part 1 we will explore the key principles and constructs resulting from the Fourth Industrial Revolution, more commonly known as Industry 4.0 as I will refer to it in this book. Part 2 will explore the emanating opportunities and challenges, illustrating the dynamics of capturing value from innovation in manufacturing with real case studies created as part of in-depth research on the impact of Industry 4.0 implementation in manufacturing from a business model point of view. Finally, in Part 3 we will provide a value-driven framework and a set of tools and recommendations for practical action to improve the chances of creating and capturing value from the implementation of Industry 4.0 in your business.

The purpose of each chapter is to provide a practical account of the relationship between the relevant facets of Industry 4.0, manufacturing business models and the potential streams of value creation, to reflect on the business aspects (as supposed to only operational improvement perspectives) of these relationships and to identify potential future challenges.

The emergence of Industry 4.0

During the course of the last two decades, Industry 4.0 has evolved from a vague and misunderstood term to a concept widely discussed and recognized by industry, academia and government. The concept encompasses the systematic digitalization of individual businesses, as well as the connection of multiple businesses within and across industries encompassing an entire value chain in real time.

A value chain for the purposes of this book is defined with the same business management perspective as originally proposed in 1985 by Michael Porter, the world-leading academic and Harvard professor[2]. According to him, a value chain is a set of activities performed by various companies operating in a particular industry in order to create and deliver value to its customers.

This concept of the value chain is based on the idea of a logical perspective presenting the business as a system consisting of a number of sub-systems each with its own inputs, value-add processes and outputs, all of which involve the acquisition and utilization of tangible and intangible resources, such as funds, equipment, materials, labour and knowledge.

A fundamental aspect of Industry 4.0 is the connectivity and the availability of all relevant information in real time through the entire value chain, encompassing all businesses and processes involved in value creation (product design and development, manufacturing planning, manufacturing operations, sales and services).

This real-time set of data opens up the opportunity to simulate alternative scenarios, optimize the value chain, and derive the best decisions and possible value streams from data at all times. As a result, the manufacturing value chain is undergoing a phenomenal transformation that affects all businesses involved.

This connection of various businesses across the value chain affects both existing and new relationships in industry. These relationships are at the core of how manufacturers interact with their customers and suppliers and, more importantly, how they generate and capture value.

On a practical level, as this book will show, despite the change in innovation paradigms evident in manufacturing since the early 1990s, most manufacturers still operate business models solely based on revenue derived from tangible assets (sales of physical products) and only a small minority is aware of the opportunities related to business models where intangible assets (these include all forms of IP) are used as a source of revenue or at least as a source of competitive advantage to achieve higher margins on sales of tangible assets.

The disregard to IP as a source of value in manufacturing is widely recognized across industry. According to the Manufacturing Technologies Association (MTA)[3], each year millions of pounds in IP rights are being neglected by the UK's manufacturing and engineering industries, because companies do not understand the value of intangible assets within their businesses, designs, products and processes.

This issue increases in importance with the integration of the value chain, as manufacturers who are making substantial investments to innovate their businesses by implementing Industry 4.0 technologies are exposed to the potential loss of intangible assets to competitors due to their lack of IP awareness and the increased information exchange in the digital environment. This in turn results in the failure to protect and appropriate value from their innovations.

This issue was highlighted in a report by Intel's Accelerate Industrial team in 2019[4], which provided a comprehensive view of the digital transformation of the manufacturing sector through a study that interviewed over 400 manufacturing businesses. It found that IP privacy, ownership and management is the second biggest threat affecting return on investment from digital transformation according to the manufacturers interviewed. IP challenges are second only to the challenges related to the technical skills gap, which in my view has a direct correlation to the lack of awareness regarding the value of intangible assets in the digitalized world.

Understanding this change in how value is generated and captured in digitalized value chains is a key factor to enable businesses adopting Industry 4.0 technologies to formulate appropriate strategies for digital transformation and, as a result, to profit from Industry 4.0.

To address this particular aspect of Industry 4.0 transformation, this book will explore manufacturers' current and future business models, which for the purposes of this work are defined as 'a representation of a firm's underlying core logic and strategic choices for creating and capturing value' (Shafer, Smith, and Linder 2005: 202)[5].

We will be taking a journey through what is Industry 4.0? how will it affect the creation, capture and delivery of value in manufacturing? and what are the most effective strategies for value protection utilizing adequate IP strategies?

The book is based on an empirical research study, which critically analysed how digital transformation affects manufacturing businesses models by conducting 31 in-depth interviews with senior managers and engineers across 15 organizations across the manufacturing value chain in the UK. In addition to the interviews, this book also contains insights into the contractual agreements governing the relationships between these organizations.

In total, eleven contractual agreements were analysed. These sprang from four case studies that explored the impact of different levels of digital transformation across product and process technologies, highlighting the opportunities and challenges for creating and capturing current and future value.

The empirical findings from these case studies expose five changes with a profound impact on manufacturing businesses:

- Digitalization leads to a shift in the nature of knowledge in manufacturing towards end-to-end codification of tacit knowledge.
- Manufacturing industry is in a pre-paradigmatic phase where uncertainty regarding the impact of digital technology utilization is the norm.
- There is evidence and recognition of a shift from tangible (physical) to intangible (intellectual) assets as a source of competitive advantage between manufacturers in the value chain.
- There is also evidence of limited knowledge and application of protection mechanisms for value appropriation.
- There is strong evidence that manufacturing businesses lack awareness of how digitalization changes the value chain in general.

This particular emphasis on value creation and capture is paramount as, despite the large amounts of investment in the digitalization of manufacturing evident in a variety of reports (McKinsey, 2016[6]; *The Manufacturer*, 2017[7]), there is still limited understanding as to how value from this investment is to be realized, where this value will be generated and how it will be captured by manufacturing businesses.

This uncertainty about the landscape of economic benefits can perhaps be attributed to the fact that so far much of the focus for investment and research has been placed on initiatives with the objective of making advances in operational and technological aspects of Industry 4.0, leaving a gap in understanding regarding the impact on businesses and business models (ie, how value is identified, generated, distributed and captured).

In this context, Industry 4.0 is having uncertain effects that lead to a highly dynamic environment. The disruptive nature of the combined power of technologies will lead to many challenges and opportunities regarding the strategic and operational aspects of a business. Mastering and overcoming such challenges will lead to a competitive advantage in this new industrial paradigm.

Manufacturing's additional value

Manufacturing plays a significant role in the UK economy. As measured in the national accounts, it provides over 2.7 million jobs, makes up 49 percent of UK exports, and contributes 66 percent of all UK business expenditure on research and development. However, manufacturing's contribution to the UK economy, about 9 percent of gross domestic product, may seem dwarfed by services, which make up 70 percent.

However, it has been continuously argued that the economic value of manufacturing is being underestimated in official statistics. According to a 2019 report by the Institute for Manufacturing at the University of Cambridge[8], as much as half of the total real value of manufacturing is missed, causing significant issues for policymakers.

An implication of this finding is that, if the way manufacturing-related activities are accounted for does not change, the UK could be missing significant opportunities to build world-leading industries. It is also critical that post-Brexit international trade negotiators are equipped with a more accurate understanding of the value of these industries and in particular the potential economic impact of companies moving manufacturing operations away from the UK.

I am in agreement with the Cambridge report, as in my view the manufacturing impact may in fact be significantly higher in economic contribution and underestimating it could have serious implications for national decision-making.

The official manufacturing figures and statistics do not account for the additional value generated through jobs and services provided across the value chains. This value added should be credited to UK-based manufacturing, as many of these jobs and services would not exist if not to support manufacturers.

In fact, many of these services, in particular, technical and professional ones, require deep knowledge and sophisticated capabilities related to the manufacturing activities they support. This argument demonstrates that the traditional view of manufacturing as a tangible asset is embedded in the fabric of our economy and, in particular, in the way we measure industry.

Throughout this book, I will seek to demonstrate new businesses models and new sources of value for manufacturing enabled by Industry 4.0. Nevertheless, in order to benefit from these you will have to understand manufacturing as an activity the goes way beyond the production of physical products. Rather, it is an activity that involves continuous innovation and technology development, creating valuable intangible assets which could be used to help your business capture value and secure your position in the highly competitive markets in which you operate.

Intellectual property and digital transformation

Perhaps this traditional emphasis on manufacturing business models solely based on sales of tangible assets comes from the conventional wisdom that IP strategy is about 'the sword and the shield'. As a sword, IP was used to attack a competitor who seeks to exploit some aspect of your IP in a way that violates your rights. As a shield, IP can help you to defend yourself from attacks by competitors.

This view of IP still holds partially true today, but this battlefield metaphor, which suggests control of IP to the greatest extent allowed by law in every instance regardless of context, is well out of date. In the context of Industry 4.0 and the new businesses models it enables, this traditional view of IP strategy is not representative of the best approach to profit from digital transformation and can actually hinder your business when it comes to collaborative relationships across connected value chains.

In this book, I argue that IP is better represented by images of collaboration, innovation and deal-making than battlefields. Courtrooms are only a last resort if something goes terribly wrong. Your business is less likely to benefit from an exclusively sword-and-shield strategy with respect to IP. However, if you are like many engineers and managers in manufacturing, you are probably not paying much attention to your business's IP and consequently your organization can be exposed to risks emanating from the implementation of Industry 4.0, notably from the exposure of sharing data across the value chain.

Furthermore, I propose that to succeed in the era of Industry 4.0 connected value chains, you and your business should dedicate greater attention to IP strategy in order to support the capture of returns on R&D investment, particularly in regard to digital transformation. This argument is not only relevant to those who run high-tech businesses. Industry wide, the traditional way of thinking about IP, or even completely dismissing it, will have a detrimental and limiting impact on your business, which in turn, will lead to short-sighted decisions.

A great deal of creativity and agility are necessary for a profitable long-term IP strategy. The strategic positions grounded in openness and collaboration, like those required to implement and profit from Industry 4.0, will offer significant benefits to your business if you are willing consider new approaches and keep an eye on the long-term game.

Manufacturers that use IP as a flexible asset class capable of supporting the organization in a broad range of ways and in formulating the overall businesses strategy will be in a better position to succeed in their Industry 4.0 journeys. This book will provide you with actionable insights for special consideration in regard to strategies for openness rather than exclusion. This advice holds true independently of the size or type of your business, whether you are a well-established and large company, an SME or even a new start-up.

If you take a bird's eye perspective, IP is a way to characterize your business capabilities, your team know-how and what your business can do. In this manner, IP represents the collective and accumulated knowledge, skills set and outputs of all the individuals making up each of the departments which make up your business. If you consider the management of knowledge and capabilities as part of your business's overall strategy, you will surely get this perspective of IP intuitively.

IP can also be viewed as a critical and flexible asset class that can be leveraged to support your business in achieving a wide range of goals: from supporting access to new markets, to improving and enhancing products, to enabling new revenue streams. A flexible asset class means that IP can represent more than just a line item on your intangible asset register or in your balance sheet, it can be used in a wide range of ways to help achieve your business's mission in the long term. As an asset class, IP represents an important set of resources available to support any business in achieving its overall mission, which has intrinsic value with potential to be transformed into a competitive advantage in existing and future markets.

In many businesses, IP is already considered a strategic asset. The IP licensing market generated over $290 billion globally in 2019. Businesses based in the United States and Canada earned approximately $170 billion in annual revenue (Licensing International, 2020[9]).

Even if this huge, and growing, market size does not get your attention, you should consider the fact that over the last 40 years, intangibles have evolved from a supporting asset into a major draw for investors. Today, they make up 84 percent of all enterprise value on the S&P 500, a massive increase from just 17 percent in 1975 (*Visual Capitalist*, 2020[10]).

The key takeaway from this book is simple: if you want to profit from Industry 4.0, your business will need to consider carefully what is your most valuable IP and how you use it in the context of your current and future relationships across the value chain. This will require developing an understanding of what intangible assets your business owns and how these assets are impacted by your existing contracts, relationships, digital transformation plans and, in particular, how your business creates and shares data across the value chain.

Whether or not your business is already making good use of your IP, there are many opportunities to increase your revenue and strengthen your competitiveness through the use of intangible assets in support of your business goals. Setting a robust strategy to leverage your existing and future IP through licensing, for example, might require careful identification and valuation of your business's intangible assets in a new way. This strategy might involve licensing what you already have in your IP portfolio. It might also require the acquisition or development of IP one way or another, or it might even mean giving away some of your IP for the benefit of your long-term strategic plans.

The simple argument that runs through this book is that you should consider each of these approaches as part of your toolbox used in the context of strategies for profiting from Industry 4.0 adoption. Whether your business likes it or not, the world is getting increasingly connected and to benefit from future opportunities to grow you should favour strategies of relative openness as the implementation of Industry 4.0 will require increasingly open relationships across the value chain with your customers, your suppliers and even your competitors.

Notes

1. World Economic Forum (2016), *The Fourth Industrial Revolution: What it means, how to respond.*
2. Porter, M. E. (1985 and republished with a new introduction, 1998), *The Competitive Advantage: Creating and sustaining superior performance*, Free Press, NY
3. 'Industry 4.0 UK Readiness Report', a white paper published by *The Manufacturer* (2017), available at: https://www.themanufacturer.com/reports-whitepapers/the-manufacturer-industry-4-0-uk-readiness-report/
4. Intel (2019), 'Creating lasting value in the age of AI+IoT: Futureproofing your business', available at https://newsroom.intel.com/wp-content/uploads/sites/11/2019/12/futureproofing-your-business.pdf.
5. Shafer, S. M., Smith, H. J., & Linder, J. C. (2005) 'The power of business models', *Business Horizons* 48(3), 199–207.
6. McKinsey Digital (2016), *Industry 4.0 after the initial hype*, available at: https://www.mckinsey.com/~/media/mckinsey/business%20functions/mckinsey%20digital/our%20insights/getting%20the%20most%20out%20of%20industry%204%200/mckinsey_industry_40_2016.ashx
7. 'Industry 4.0 UK Readiness Report', a white paper published by *The Manufacturer* (2017), available at: https://www.themanufacturer.com/reports-whitepapers/the-manufacturer-industry-4-0-uk-readiness-report/
8. Report by Jostein Hauge and Eoin O'Sullivan (2019) in *Inside the Black Box of Manufacturing: Conceptualizing and counting manufacturing in the economy,* Institute for Manufacturing, University of Cambridge, available at: https://www.ifm.eng.cam.ac.uk/uploads/Research/CSTI/Inside_the_Black_Box_of_Manufacturing_report_FINAL_120619.pdf
9. Licensing International (2020), report available at https://licensinginternational.org/get-survey/
10. Report in *Visual Capitalist* (2020), available at https://www.visualcapitalist.com/intangible-assets-driver-company-value/

PART 1

INDUSTRY 4.0 AND THE IMPACT ON MANUFACTURING BUSINESS MODELS

1.
THE FOURTH INDUSTRIAL REVOLUTION

Industry 4.0 first emerged out of German manufacturing as a result of a study commissioned by the German government in 2008. Since then the idea has spread widely and has been adopted by almost every other industrial nation. The term Industry 4.0 is a reference to the Fourth Industrial Revolution with the first three taking place in the form of the introduction of steam power in 1784, electric power in 1913 and automation in 1969.

The Fourth Industrial Revolution resulted from the introduction of cyber-physical systems (CPS), which enabled by the internet of things and the internet of services, are becoming integrated with the manufacturing environment. As suggested by Klaus Schwab, executive chair of the World Economic Forum, Industry 4.0 is going to change everything in our lives just like the previous three. However, in contrast to the previous revolutions where all the benefits and changes were only noticed later, we have a chance to proactively navigate and influence the way this revolution shapes and transforms our world.

Industry 4.0 will enable manufacturers to build global networks and to connect every aspect of their value chains including machines, production lines, factories and warehouses via CPS. In turn, by connecting and controlling each part of the value chain intelligently, data can be collected and shared to support decision-making and optimize actions.

These CPS will be configured in the form of intelligent organizations, commonly referred to as smart factories, machines or supply chains. These smart organizations will provide many opportunities for improvement in industrial processes as a whole through engineering, manufacturing, supply chain, distribution, retail and product lifecycles. This horizontal value chain is where Industry 4.0 has the potential to provide the biggest value to manufacturers.

Smart factories

One of the key building blocks of horizontal value chains is the creation of smart factories. These will transform the way manufacturing operates, using machines with smart capabilities to make smart products with the built-in intelligence to self-identify and locate at all times during design, manufacture, use and maintenance.

Technologies, such as radio frequency identification tags, allow smart products to know what they are, how they are made, what components they require, how they were manufactured and their current state. This level of intelligence requires products to be aware of their context and have access to their own make-up, production history and process for becoming a finished product. Such knowledge will let them self-select production process routes to ensure the most effective use of facilities. For example, products will be able to communicate with the production line and alter the route of the conveyor belt, re-directing themselves to a different process or even another production line.

Vertical integration

This interaction with production lines results from another concept emerging from Industry 4.0: the vertical integration of manufacturing processes. As a term, it refers to the connection of business processes within a single organization (sales, engineering, production, logistics and finance among others) and their respective information technology (IT) and operational technology (OT) systems. This integration within an organization is what will provide factories with the ability to control the end-to-end management of entire manufacturing business models from product design, supply chain management, manufacturing operations to lifecycle management.

The integration of IT and OT is a particular challenge commonly faced by manufacturers hoping to benefit from Industry 4.0. One of their main challenges is the current skills and capability gap between IT and OT professionals who tend to operate in their own silos. However, in the Industry 4.0 age these disciplines will have to merge and act as one.

Vertically integrated smart factories do not just apply to large organizations. In fact, they are ideal for enhancing the value that small and medium-sized enterprises can offer. By giving them more agility and flexibility, they support the inherent capability of SMEs to adapt more quickly to customer demand.

One example of the benefits of dynamic process control and flexibility are rapid, last-minute design changes in response to demands for the customization of products. Such integration allows for the efficient production of small lot sizes in a customized, yet cost-efficient and profitable manner.

Such applications of Industry 4.0 open up new ways for all manufacturers, including SMEs, to create value and develop innovative business models. In summary, to achieve the vision of Industry 4.0 and all its potential benefits, organizations and entire value chains will require a high degree of integration, which will bring new challenges in how future value, business models and downstream services are created.

Potential benefits of investing in Industry 4.0

The following section will provide a non-exhaustive list of typical benefits associated with the implementation of Industry 4.0 across the entire value chain.

Increased business competitiveness

Industry 4.0 has the potential to improve the level playing field through co-operation and wider collaboration between manufactures. This in turn can improve manufacturing competitiveness and create a level playing field in areas such as labour costs. For example, lower technology costs and diffusion of knowledge will result in new manufacturing models that allow developed countries in Europe to challenge the cost models for low labour cost countries in Asia, particularly when the cost of transport and the carbon footprint are taken into account.

Industry 4.0 will also enable SMEs to work together to pool their resources and capabilities to challenge large companies. It is expected that within the next decade a large proportion of consumer products will no longer be built by low-cost workers in Asia, but rather developed and made by engineers, workers and programmers as close to the end customer as possible to limit the environmental and social impacts.

Higher levels of productivity

Digitalized and connected engineering processes (from product design, production planning, product manufacturing, use in the field, maintenance and end-of-life planning) will lead to improvements in speed to market and operational productivity, as well as reductions in engineering and manufacturing costs, resulting in higher profitability and improved overall manufacturing productivity.

Increased revenue

Industry 4.0 and digital transformation will enable manufacturing as a whole to substantially grow its revenues, even though the implementation of these technologies requires significant investment. The returns on these investments can be high, however predictions are often too optimistic, as not all companies are recovering the cost of their investments, as we shall see in Chapter 5.

Higher demand for workers

With the introduction of new technologies, the requirement for new jobs and employment rates will increase, especially in areas such as engineering, data science and IT. The other side of this transformation it that there will be a level of job displacement in regard to traditional labour workers who will need to be re-trained or lose their jobs. This change is likely to result in a small net gain on jobs: more will be created than lost.

However, the job losses are likely to go far beyond just low-skilled labour and include any jobs that can potentially be performed more efficiently by an AI solution or an IT service. For example, local IT specialized jobs, such as network engineers, are likely be replaced by remote systems based on immersive technologies to support troubleshooting remotely. On the other hand, new employment opportunities will not be limited to programmers and data scientists; there will always be work for industrial process analysts and for supervisors to watch over the integrity of infrastructure, product lines and machines.

Optimized manufacturing processes

Integrating IT and OT systems is always a complex challenge that a limited number of companies have succeeded in executing. However, merging the IT and OT systems will enable manufacturers to increase efficiencies across the business. Managers and system administrators will be able to control and streamline processes in nearly real time. This will also enable increased levels of collaboration between manufacturers, suppliers and other partners in the manufacturing value chain. This will shorten the lead time for product design and production design, making the process more efficient and simplifying engineering relationships across the value chains without affecting quality.

A platform for the development of technologies

Industry 4.0 will form a platform for further innovation, as suppliers, manufacturers and technology providers use integrated systems to develop and test future solutions. It's a similar to how mobile phone technologies and applications have been developed with open application programming interfaces (APIs). In the context of Industry 4.0 we have already seen an increase in collaborative efforts to develop solutions to improve upon current technologies for sensors, controllers and measurement. Their use will enable accurate digital twins of products and processes which in turn support the development of novel materials and processes.

Improved customer understanding and better service

Industry 4.0 will support structured closed information loops to monitor and act upon direct as-used and as-maintained product feedback enabling near real-time product and process improvement. This will support concepts such as closed loop engineering where product-use information is captured, analysed and fed back to design and manufacturing engineers so that they can optimize the next iteration of the products and services based on their understanding of their user. Similar approaches are in use in supply chain and logistics where dashboards report in close to real time.

Whether you like it or not, Industry 4.0 is here and will bring changes to every aspect of your business, including business models, and the key elements of customer relationships, customer channels, value propositions, resources, suppliers, cost and revenue models.

The remaining chapters in Part 1 will explore the definition of Industry 4.0, how Industry 4.0 will affect manufacturing business models (the key bottleneck to transforming and benefitting from Industry 4.0) and, finally, why some businesses are failing to capture value from Industry 4.0.

2.
DEFINING CHARACTERISTICS OF INDUSTRY 4.0

Despite the long list of potential benefits and the development of Industry 4.0 over the last decade, we still lack a universally accepted consensus when it comes to the definition of Industry 4.0. Different industries have distinct use cases to refer to what Industry 4.0 is or is not. Definitions are still evolving.

Nevertheless, in this chapter we will discuss four commonly accepted features that distinguish Industry 4.0 from other revolutions. But first let's look at a few Industry 4.0 definitions:

> The term Industry 4.0 stands for the fourth industrial revolution. Best understood as a new level of organization and control over the entire value chain of the lifecycle of products, it is geared towards increasingly individualized customer requirements (Rüßmann et al, 2015[1]).

> The concept encompasses the digitalization of individual businesses and the connection of businesses, within and across industries encompassing an entire value network in real time (*Plattform Industrie 4.0*, 2013[2]).

The range of definitions available in academic and practitioner publications represents various vantage points and different levels of abstraction. For example, Faller and Feldmüller (2015)[3] focus on a narrow definition with an IT perspective on Industry 4.0, defining it as: 'IT integration of the production level with the planning level and further on to customers and suppliers'.

Other authors offer a view of Industry 4.0 as a more broadly based set of changes in the market place. to customer needs, within organizational and hierarchical developments, and as new working methods (Lasi, Fettke and Kemper 2014[4]; Magruk 2016[5]). The definition offered by Kirazli and Hormann (2015)[6] focused on a wider aspect and states that: 'Industry 4.0 is the systematic development of an intelligent, real-time capable, horizontal and vertical networking of humans, objects and systems'.

On the other hand, Wang et al (2016)[7] define Industry 4.0 in a narrower sense as the interconnecting of a production system, linking together various associated functions, such as logistics and warehousing, but within a single business unit.

Definitions focusing primarily on IT transformation or an overhaul of manufacturing and associated business models also vary, arguably pointing to the fact that there is still a high degree of uncertainty as to what Industry 4.0 really means and what are the likely implications for manufacturing businesses (Almada-Lobo, 2015)[8]. Some authors such as Oesterreich and Teuteberg (2016) offer a definition covering a number of factors and concepts: 'multifaceted term comprising a variety of interdisciplinary concepts without a clear distinction' (Oesterreich and Teuteberg, 2016, 122)[9].

Some of these key concepts are also referred to in the literature as the technology enablers or key enablers, including big data, 3D printing and robotics (Almada-Lobo 2015). The majority of definitions found in the literature focus on the operational impact on production and manufacturing processes. However, there are also references to the impact on the wider organization of business, as in the definition offered by Schuh (2014)[10] who identifies that the effects of Industry 4.0 will also impact indirect business functions.

Finding a consensus regarding the meaning of Industry 4.0 can be difficult; people in practice tend to define it as 'making manufacturing industry fully computerized' or 'making industrial production virtualized'. A level of cohesion appears in the form of an agreement that Industry 4.0 'integrates horizontal and vertical channels'. Despite this challenge in defining and giving practical examples of Industry 4.0, manufacturers remain keen to keep up with the rapid pace of change and hope to benefit from deploying the related technologies.

To offer some clarity, Industry 4.0 can be seen as essentially a process of digital transformation leading to an improved approach to efficient manufacturing, which utilizes a complex combination of the latest technological advances, particularly those that enable the integration of physical and digital assets, and the merging of IT and OT.

The Institute for Digital Engineering (IDE) has succinctly presented the distinction between various levels of digital maturity in the *Digitalization Roadmap* published in early 2021 where it is emphasized that the level of digital transformation in Industry 4.0 requires holistic integration of digital technologies into all areas of business enabling fundamental changes in business models, enterprise architecture and culture (IDE, 2021[11]).

The key objective here is to connect and manage the whole value chain process, improve efficiency levels across the entire product lifecycle, and develop novel products and services that generate more value to customers through better quality and higher customizability, but not at the expense of lower costs.

Four common features of Industry 4.0

Despite the lack of a clear definition, the proponents of Industry 4.0 have been consistent since its conception in identifying four distinct characteristics at its core:

Vertical integration

In order to create the smart factories, which are at the centre of Industry 4.0, smart technologies have to be deployed in a co-ordinated and purposeful manner. The smart factories must be connected in a coherent way to create a link between the products, the processes, the machines, the production lines, the factory and the business. At the core, vertical integration in this context is used to refer to cyber-physical systems, which connect an entire factory and enable maximum efficiency and flexibility to react quickly to customer requirements and unexpected disruption.

Horizontal integration

The next level of integration is aimed at creating a network of vertically integrated and optimized businesses that enables efficient value creation across the value chain. An example of a horizontally integrated value chain that springs to mind is one where a car factory is connected to its customers and suppliers. The customer uses an online vehicle configuration tool which allows the selection of customized options for the vehicle. Upon the configuration of the vehicle, the customized options are converted into product and process requirements which are distributed in near real time across the value chain for the efficient production and assembly of the car. This level of integration also results in a number of opportunities for the creation of new business models across industries and countries.

End-to-end engineering

The term end-to-end engineering, sometimes referred to in manufacturing as 'cradle to grave', is used in this context to describe engineering across the entire product lifecycle: from initial product concept, product development and production development, all the way to manufacturing, customer use, maintenance and end-of-life activities, such as repairing, repurposing or recycling. This requires the whole engineering process to be connected, often across all those organizations involved.

One example of end-to-end engineering in action is found in how new battery packs for electric vehicles are designed with the entire lifecycle and, in some cases, even with the second life in mind. This may involve the integration of an engineering house with expertise in battery design, the car manufacturer, the battery manufacturer and the recycling business to work together in designing the life of the product. So batteries are designed to have a second life, for example, after a useful life in a vehicle as part of a system for power storage for solar-powered homes.

Rapid growth in manufacturing

The three characteristics mentioned above are associated with the fourth which is the increase in pace of growth in manufacturing. This growth is realized by the improved speed of development and efficiency in engineering and operations across the value chain.

Industry 4.0: why now?

The phenomenon known as 'the rise of the machine' was popular almost half a century ago as a potential solution to process variation by humans in manufacturing. It sparked concerns about scenarios where automated machines and robotic systems would take over and completely replace humans in manufacturing.

The approach had some initial success and resulted in the establishment of automation in industry. Some experts forecast the end of manual labour within five to ten years. However, as we know well, automation did not work out quite that way and humans remain a critical part of manufacturing.

Industry 4.0 brings the next gear change in manufacturing through the digital transformation of value chains interconnecting the physical, digital and virtual worlds. This change presents new opportunities, but it requires new thinking in terms of value creation and business models. The following advances in business practices can be associated with the emergence of Industry 4.0:

- **Data processing**: the development and exponential increase over the last decade in data volumes, cloud storage, cloud computing, computer power, and ubiquitous device and network connectivity has enabled collection, processing and analysis of data at a scale that was previously impossible.
- **An exponential growth in analytics**: engineering processes for product and process development require complex models, simulation and extensive analysis. Traditionally, with better models and analysis, an organization can improve the confidence levels in the performance of a product, process or service. The same complex models and analysis are also required to improve business operations and entire supply chains in order, for example, to accelerate a product's time to market, and reduce its stock levels and obsolescence.
- **Human-machine interfaces**: these technologies include many forms of immersive technology such as mixed, virtual and augmented-reality technologies that make full use of touch interfaces, hands-free and other human-enhancing systems.
- **Data-transfer technology to support executable and physical outputs**: for example, the use of simple interfaces to link computer-aided design and engineering tools with robotics, additive manufacturing technologies and rapid prototyping technologies that contribute to a faster pace of development and higher customization potential.

An unprecedented scale of technological innovation and industrial transformation has enabled these key advances in the way we do business and commerce today. In particular, a number of key technological trends are still maturing and shaping the implementation of the new manufacturing paradigm.

Nine key technology trends

Industry 4.0 represents the confluence of numerous technologies and trends. These nine represent those that recur most often in academic and industrial spheres in the context of manufacturing digitalization.

Big data

Nowadays engineering and manufacturing are data-rich environments. Data sets are being created, processed, stored and analysed at unprecedented levels of speed and size. These data sets are generated at various sources, but all the data is usually collated and organized in a coherent manner in order to support decision-making across businesses and value chains. Your business cannot afford to ignore the data coming in, as it might prove to be useful when it comes to optimization of your production quality and service, reduce energy consumption, and improve efficiencies in the production process.

For example, data collected from the various phases of the production process can be analysed in correlation with each other to identify phases where performance falls below the desired standards or where activities may be optimized. In summary, we could say that six Cs exist in the context of big data in the Industry 4.0 manufacturing ecosystem:

- Connectivity of sensors and networks across your factory
- Cloud-computing solutions
- Cyber-physical systems
- Content or context awareness
- Collaboration or co-operation across the value chain
- Customization

Autonomous robots

The deployment of robotics is nothing new in the manufacturing value chain. Your business will surely have used and benefited from process automation and robotics in the past. Nevertheless, robotics has also been subject to accelerated development and evolution. The new generation of autonomous robotic systems are designed to be self-aware, self-sufficient, autonomous, and interactive, so that they are no longer simply tools used in your factory to perform standardized, repetitive and controlled tasks, rather they are already integral work units that function in collaboration with humans in operational and in back-office functions as in the case of collaborative robots (cobots) and robotic process automation (RPA).

Simulation

In the past, if your business wished to test the efficacy and efficiency of a given product or process, a great deal of effort, experience, and trial and error was required. On the other hand, Industry 4.0 uses virtualization to create digital twins that are used for simulation modelling and testing. These digital representations of physical observable objects will play a major role in the optimization of production, as well as product quality and product uses in the field.

Digital twins are often complex, but they can be defined generally as 'digital representations of a physical asset, sufficient to meet the requirements of a set of use cases for which it was created'. According to ISO 23247, which sets the standard for digital twins in manufacturing, a digital twin consists of three elementary aspects: data, models and service interfaces. The level of abstraction and complexity of a digital twin must be such that it is sufficient to meet the requirements of the use cases for which it is designed.

Horizontal and vertical system integration

Industry 4.0 aims to integrate IT and OT systems and environments to create a scenario where engineering, production, marketing, distribution and after-sales are connected across all stakeholders. As a result, business across the value chain will be integrated, resulting in the creation of integrated networks, collaborative innovation and new levels of collaborative automation, and even value chains that are autonomous and self-optimizing.

The industrial internet of things

The connection via the internet of computing devices embedded in objects, enabling them to send and receive data is known as the internet of things. In industrial sectors, its applications are being extended to machine-to-machine (M2M) communication and big data. IIoT is an essential element of Industry 4.0 based in the use of transducers and devices equipped with low-power radio networking to allow the entire network to communicate and interact, while connected to a network gateway for its control and management. This architecture for the connectivity of devices will become ubiquitous throughout the smart factory and Industry 4.0 connected value chains.

Cybersecurity

With the integration of OT and IT, the connectivity of the wider network of industrial systems across the manufacturing value chain is increasingly vulnerable to external threats. This can be seen by recent attacks on industrial targets, especially manufacturers adopting new digital technologies[12]. To mitigate risks associated with this new source of vulnerability, digitalized manufacturing businesses will have to put in place cybersecurity measures that recognize the new challenges of integrated OT, IT and IIOT environments.

The cloud

As mentioned under big data, Industry 4.0 involves large amounts of data which require collection, processing and analysis in order to leverage the potential benefits of digital manufacturing across the value chain. One of the problems with these large volume of data is that not many manufacturer have the infrastructure required to store and analyse the data collected. This is where cloud-based solutions can provide additional capabilities and capacity on demand at the point of use to support business to create private clouds suited to their storage and processing needs for manufacturing data.

Additive manufacturing

AM, such as 3D printing, has been around for a long time. However, the reduction in cost and technological improvements over the last two decades has democratized the technology and enabled even small manufacturers to benefit from the creation of on-site rapid prototypes and proof-of-concept designs. In turn, this has reduced time and effort in product and process design. The technology also allows for the production of small batches of customized products, offering value to customers, while reducing risk, capital cost and time inefficiencies for the manufacturer.

Immersive technologies

Immersive technologies such as augmented, mixed and virtual reality have a number of use cases in the Industry 4.0 context. These include the ability to reduce training and maintenance costs, as well as the overheads associated with design, manufacturing, marketing and after-sales support. Manufacturers are already using immersive technology solutions to enhance their manual operations for complex products, conducting training and supporting maintenance, while reducing the costs of having dedicated local experts onsite.

These nine technology trends continually push manufacturers to keep the pace of innovation up, especially if they seek to develop and maintain sustainable competitive positions. As evident from the discussion thus far in Chapters 1 and 2, there is a clear emphasis in the value creation and capture across the entire value chain. This is a key concept to understand if you are to appreciate the potential opportunities and challenges associated with Industry 4.0. This is the topic of discussion in our next chapter.

Notes

1. M. Rüßmann, M. Lorenz, P. Gerbert, M. Waldner, Industry 4.0 (April 09, 2015), *The future of productivity and growth in manufacturing industries*, pp 1–14.

2. *Plattform Industrie 4.0* (2013), Whitepaper FuE Themen, available at: http://www.plattform-i40.de/sites/default/files/Whitepaper_Forschung%20Stand%203.%20April%202014_0.pdf

3. Faller, C., and Feldmüller, D. (2015), 'Industry 4.0 learning factory for regional SMEs'., CIRP proceedings, pp 88–91.

4. Lasi, H., Fettke, P., Kemper, H.G., Feld, T., and Hoffmann, M. (2014), *Industry 4.0: business and information systems engineering*, 6 (4), pp 239–242.

5. Magruk, A. (2016), 'Uncertainty in the sphere of Industry 4.0: potential areas to research', *Business, Management and Education*, 14 (2), pp 275–291.

6. Kirazli, A., and Hormann, R. (2015), 'A conceptual approach for identifying Industrie 4.0 application scenarios', proceedings of the 2015 Industrial and Systems Engineering Research Conference.

7. Wang, S., Wan, J., Zhang, D., Li, D., and Zhang, C. (2016), 'Towards smart factory for Industry 4.0: a self-organized multi-agent system with big data based feedback and coordination', *Computer Networks*, 101, 158-168.

8. Almada-Lobo, F. (2015), 'The Industry 4.0 revolution and the future of manufacturing execution systems (MES): Letter from industry', *Journal of Innovation Management* [online] 3(4), 16–21, http://hdl.handle.net/10216/81805.

9. Oesterreich, T. D., and Teuteberg, F. (2016), 'Understanding the implications of digitization and automation in the context of Industry 4.0: A triangulation approach and elements of a research agenda for the construction industry', *Computers in Industry*, 83.

10. Schuh, G. (2014), 'Collaboration mechanisms to increase productivity in the context of Industrie 4.0', CIRP proceedings, 19, pp 51–56. Available at: www.sciencedirect.com, (accessed 18 August 2017).

11. IDE (2021), *Digitalization Roadmap*, available at: https://roadmap.ide.uk/.

12. Article available at https://securityboulevard.com/2018/08/manufacturing-a-rising-target-for-cybercriminals/.

3.
INDUSTRY 4.0 VALUE CHAINS

The original concept of the value chain, as defined in 1985 by Michael Porter in *Competitive Advantage: Creating and sustaining superior performance*, emphasized that:

> Competitive advantage cannot be understood by looking at a firm as a whole. It stems from the many discrete activities a firm performs in designing, producing, marketing, delivering, and supporting its product.[1]

In other words, you create competitive advantage by going beyond improvements at the operational level and maximizing value across the entire set of processes involved in developing and commercializing your product or service.

It is no surprise that most companies seek to optimize their value chains, as they need partners across the lifecycle from research, design, development, marketing and manufacturing. Therefore, a value chain requires that businesses work with partners that have skills in certain disciplines in order to enhance their value to their current and future customers. If done right, it can create long-lasting, sustainable competitive advantage.

Porter's definition provides two categories of business activities that form part of all value chains: primary and supporting activities.

Primary activities

Primary activities generally consist of five processes, which are essential for adding value and creating a competitive advantage:

- Inbound logistics: these processes include functions like receiving, warehousing and managing inventory.
- Operations: these processes include activities to convert raw materials into finished goods.
- Outbound logistics: these activities include processes related to the distribution of a finished product to a customer.
- Marketing and sales: these activities include setting out strategies to enhance visibility and target appropriate customers.
- Service: these include programmes to support and maintain products and enhance the consumer experience, including customer services, maintenance, warranty and repair.

Support activities

Four support activities generally complement these primary activities. Increases in their value and efficiency typically lead to a direct benefit to at least one of the five primary activities. These support activities are generally identified as overhead costs or indirect costs in business accounts:

- Procurement activities: how a company sources raw materials for its products and services.
- Technological development: these activities concern research and development in designing and developing products and processes, such as manufacturing techniques and automating processes.
- Human resources: these activities are particularly important, as human capital is critical for supporting sustainable competitive advantage. The tasks under this heading include the hiring,

training and retaining of employees who will enable the business to delivery its vision and strategy.

- Infrastructure: these activities include company systems and the composition of its management team, encompassing planning, accounting, finance and quality control.

In order to formulate an optimized value chain, you have to consider both primary and support activities as part of your business strategy. This strategy should consider every business process and ensure that the value chain and the involved partners are aligned. Digital transformation enables the creation of versatile and agile value chains aimed at maintaining long-term sustainable competitive advantage.

As we have discussed, the characteristics of Industry 4.0 support all these functions, creating the potential for higher productivity, efficiency and self-managing production processes. In these production environments, people, machines, equipment, logistics systems and work-in-process components communicate and co-operate with each other directly. In Industry 4.0 manufacturing environments, enterprises do not place all their focus on continuous improvement of operations, ie, speeds and feeds of the machinery. Instead, they focus on how Industry 4.0 technologies, intangible assets, new value chains and new manufacturing methods can drive revenue.

A distinction is made between digitization and digitalization. The real value of Industry 4.0, particularly in manufacturing, lies in moving beyond digitization to digitalization. The distinction between these terms is not always clear. Even Industry 4.0 practitioners confuse them.

Digitization involves creating a bits-and-bytes version of analogue or physical things, such as microfilm images, paper documents, photographs and sounds. In smart factories, digitized information can create the virtual equivalents of assembly line components or even an entire factory floor.

On the other hand, digitalization takes this transformation process one significant step further. Physical things can be digitized (be represented by a digital twin) but only business operations, functions, models or processes can be digitalized. By leveraging digital technologies, it is these improvements that can deliver maximum value for the least possible total cost and create a competitive advantage.

Defining benefits as process improvements brings us directly to the idea of generating value. More and more analysts view digitalization as the road on which businesses move from process-level improvements towards digital business and digital transformation.

Business value with Industry 4.0

Industry 4.0 has changed how we view business value and the potential to increase it. A growing number of economic analysts suggest that manufacturers have already reached the end of the line for value from traditional cost-cutting measures.

Manufacturers throughout the world are eager to benefit from the business value that Industry 4.0 might generate. Successful adoption of Industry 4.0 technologies will involve looking for value in some new places, such as in customized products, services and product design. For manufacturers, it's now logical to investigate what benefits they might expect from digital transformation.

The smile curve

The concept of the smile curve was originally introduced by Acer Incorporation's founder Stan Shih[2] in the early 1990s when he used his diagram of a smile curve to emphasize how value is generated across processes in manufacturing's value chain. In this diagram, the middle of the smile, where the actual manufacturing takes place, is the lowest point of value creation in comparison to the two ends of the smile diagram, which represent the highest points of value creation at each corner, where R&D at the beginning and customer services at the end are positioned.

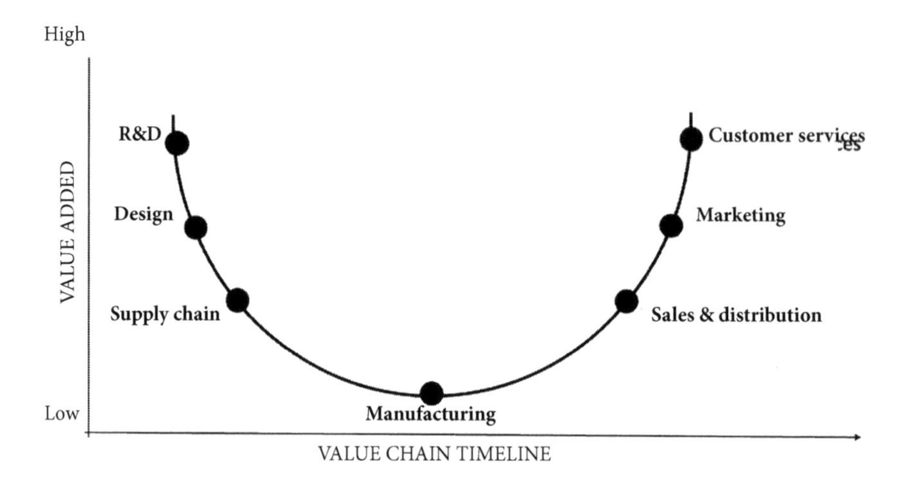

Figure 1: The smile curve of value across the value chain

This concept is widely accepted and became the foundation for an article featured by *BusinessWeek* in 2017[3], 'Factories won't bring back the American dream', where the authors analysed a study carried out by the Asian Development Bank Institute exploring the Apple iPhone's value chain. The study found that the manufacturing process in China accounted for only 3.6 percent of the iPhone production cost. The remaining 96.4 percent of the production cost was distributed across technology suppliers and Apple which receives the vast majority of the profits as the holder of the intellectual property required to make and commercialize the product).

The report also highlighted that at the time of the study the net income at Apple, which does almost no manufacturing, was an impressive 21 percent of revenue, its shares traded at 18 times earnings, and it captured 91 percent profit margins. Whilst, Taiwan's Hon Hai Precision Industry Co., one of the companies to which Apple outsources its manufacturing, recorded net profit of 3.5 percent on sales.

So, if the value to industry is not in the actual manufacturing of the products, where can Industry 4.0 help to improve value creation? For many manufacturers, making a product still involves assembling components and selling the product. The product's value lies in the selling price of the item and the efficiency of the manufacturer's operations.

Now, the focus is moving from making a product to results (outcomes), such as customer satisfaction, safety or comfort. It also involves monetizing new stages of a product's life cycle, such as R&D, customization, intangible assets, marketing, customer support and product insights. For manufacturers in many industries, the outcome-based economy is about:

- bundling products and services;
- leveraging intangible assets to secure market position;
- moving from one-time to ongoing transactions;
- enabling subscription-based transactions based on a recurring revenue relationship.

In short, the outcome economy emphasizes monetizing the entire value chain and everything about the product, production or services.

Finding the value in manufacturing

It is easy to say you and your business should start selling more than your traditional physical products or services. However, changing your business model is much more complicated. This change is a journey, which begins with business leaders learning where the value in their own manufacturing value chain lies and it is often not where they might have expected.

As mentioned in the case of Apple, the value in a manufactured product lies in the concept, R&D, branding and product design at one end and distribution, marketing, sales and services at the other. When people talk about Industry 4.0, they often mention a two-tier value regime. The first category is what people typically think about value: cost savings, improved worker productivity and faster times to market. These factors all depend on producing more products, more efficiently. Some analysts call this low-hanging fruit, which can be achieved by improving current manufacturing processes to make products with traditional business models.

Value in the second tier, however, depends on new business opportunities. This is where new and existing Industry 4.0 technologies combine to create new business models and revenue. Finding value in this new frontier of manufacturing requires an agile and skilled workforce, modern equipment, high-risk tolerance and a sharp eye for business opportunities. If your business is prepared, Industry 4.0 technologies can help create value at the production, business and market levels.

At the manufacturing-related, process-level value can be generated by process operations and maintenance within businesses in the following ways:

- Costs of manufacture, maintenance, supply and distribution reduced or avoided.
- Improved supply chain transparency.
- Lower inventory costs.
- Avoiding equipment repair and maintenance costs.

At the business-level value, the sources of potential value are distributed across several levels up the manufacturing hierarchy from processes to the enterprise. This is where you find the total value of products and services generated by a business. Examples of value generated at this level are:

- Revenue added via new products, services or business opportunities.
- Revenue recovered from productivity lost to inefficient operations.
- Higher revenue enabled by improved productivity.
- Improved production and worker productivity.
- Improvements in the quality of goods.

- Greater customer satisfaction and loyalty.
- Faster product time to market or time to service.
- Faster response to customer preferences and requirements.

These types of value are not just measurements of financial performance. They also serve as indicators of potential value, growth and business opportunity.

At the level of the value chain in manufacturing, benefits typically arise beyond the enterprise. These hard-to-quantify metrics reflect improvements in education, technological development and quality of life:

- Provide modern customers with satisfaction enabled by advanced products and services that they crave.
- Improve wages and living standards for the nation's people.
- Develop a workforce with skills in new fields such as data science and analytics, data engineering and agile manufacturing applications.
- Improve innovation in fields such as analytics and advanced production automation.
- Create a strong local demand for Industry 4.0 components (IoT, analytics etc), which helps companies build capabilities in these areas.

Many people in industry and academia have a shared misconception that Industry 4.0 will only benefit manufacturing. Although, the concept originated in manufacturing, its impact will reach far beyond factories and manufacturing industry.

The creation of connected value chains using Industry 4.0 cyber-physical systems can open the value-chain doors to new entrants, including product creators, designers, machine builders, programmers, manufacturers, suppliers and workers. One of the key challenges raised by early Industry 4.0 adopters is the lack of skilled workers who understand the holistic potential to create value beyond the traditional operational use cases.

In order to roll out Industry 4.0 across the value chain, this skill problem will need to be addressed by industry and academia. The education sector will have to step up to produce more talent equipped with the skill sets and competencies required. Software and technology developers will also have to adjust their skill sets and become more aware of the intricacies of IT systems and OT systems, such as industrial engineering systems and control systems.

To address this skills gap, the World Economic Forum revealed within its report, *The Future of Jobs 2020*[4], the top ten skills required by current and future workers.

- Complex problem-solving
- Critical thinking
- Creativity
- People management
- Co-ordinating with others
- Emotional intelligence
- Judgement and decision-making
- Service orientation
- Negotiation
- Cognitive flexibility

Developing these core skills is a key step in the journey to implementing Industry 4.0 beyond the operational level. These capabilities are a critical enabler required for your business to understand and benefit from the six design principles on which Industry 4.0 value chains depend.

Six design principles for Industry 4.0 value chains

Interoperability

This principle goes beyond the integration of manufacturing processes which follows a predefined set of steps and requires resources such as people, machines and processes that are directly involved. Interoperability, in the context of Industry 4.0, requires the conceptualization of an entire environment built upon flexible interactions and collaboration between all the component parts.

As an example, you can think of products designed along with the manufacturing processes, the machines, the controllers and the work instructions for people who will work on them. This process typically takes place across the value chain and involves multiple systems, suppliers, data sets and file formats. As such, interoperability as a design principle for your digital transformation refers to the capability of all components to connect, communicate and operate together. It is one interoperable system combining humans, factories and the relevant technologies across multiple businesses in the value chain.

Virtualization

In traditional manufacturing the monitoring of processes parameters involves the use of sensors connected to physical machines and the production line, collecting and displaying data. Virtualization as an Industry 4.0 design principle means linking this physical asset data to virtualized models to enable engineers and designers to identify and predict issues, customize the process, simulate the future performance in parallel and test changes or upgrades in complete isolation, without affecting the physical processes they have virtualized. This is also know as a 'digital twin'. The virtualized twins can be used to enhance existing processes and products, as well as design, develop and test new product and processes, even creating entire production lines and virtual value chains to evaluate and reduce the risks in advance of investing in new products and production facilities.

Decentralization

Decentralization as a design principle for Industry 4.0 represents a decision-making model spread across your business and the entire value chain. Industry 4.0 supports decentralization, particularly when it enables different systems within the factory and the value chain to make autonomous decisions based on the best possible outcome or organizational goal. For example, data and information in decentralized businesses are usually more detailed, up to date and relevant than those in centralized companies. Decentralized manufacturing allows for more flexibility and agility at the local level.

Real-time Capability

Industry 4.0 efforts are also centred on making everything in real time at each stage of production, whether collecting data, monitoring processes or giving feedback.

Service orientation

The internet of things creates potential services that others can consume. Therefore, internal and external services are still going to be required by smart factories, which is why the internet of services is such an important component of Industry 4.0.

Modularity

Flexibility is also another design principle of Industry 4.0, so that factories can easily adapt to changing circumstances and requirements. By designing and building products and systems, such as conveyor belts, that are modular and agile, the factory is flexible and can change production. Manufacturers can ensure that individual product lines can be replaced, expanded or improved with the minimum of disruption to other products or processes.

The standards for digitalization

The use of these design principles and technologies across the entire value chain, including all suppliers and SMEs, will require a reference architecture to co-ordinate different areas of expertise and methods of working. It will be a general model that each company can deploy to enable one common approach to data protocols and structures, machine, devices, and communication interfaces.

The difficulty is that each company in the partnership has a different perspective, which could be from a manufacturing, software, engineering or network angle. For example, from a manufacturing perspective, the focus will be on process and transport functions. Whereas a software perspective will consider the types of application and systems management each partner uses. These interfaces between different applications for enterprise resource planning are required to enable inter-company logistics and business management planning.

From the network side, the perspective is on devices and how they connect. The list of devices to take into consideration is vast, as they can be sensors, actuators, servers, routers, field bus, tablets or laptops. The engineer perspective is more about the product lifestyle management.

Industry 4.0 architectures will rely on the use of existing standards to connect everyone in the value chain, allowing them to participate in an autonomous or semi-autonomous way, which is part of the value of Industry 4.0.

Notes

1. Porter, M. E. (1985 and republished with a new introduction, 1998), *The Competitive Advantage: Creating and sustaining superior performance*, Free Press, NY
2. See https://www.163.com/money/article/6A1GCL9H00253G87.html
3. See https://www.bloomberg.com/news/articles/2017-06-08/factories-won-t-bring-back-the-american-dream
4. See http://www3.weforum.org/docs/WEF_Future_of_Jobs_2020.pdf

4.
THE HUMAN DIMENSION OF TRANSFORMATION

The discussion so far has been heavily focused on business, technologies and systems. However, without people (technology developers, technology users, managers, engineers, customers and suppliers, just to name a few), there is no Industry 4.0. People are the link between the technologies and solutions used in Industry 4.0 and the customers, suppliers and partners, ensuring that engineering processes, production lines and entire value chains run smoothly.

Engineers, managers, technicians, inspectors and process operators, their relationships and interactions, must be a key consideration in your digital transformation journey. For example, to be effective, managers need access to enterprise systems and to the manufacturing process via dynamic real-time dashboards that display status, trends and alerts. Do not be misled into thinking that because smart machines and systems are automating most of the standard activities and straightforward decision-making that there will be no need for people in manufacturing. Even the most advanced autonomous systems still require a human to deal with exceptions and anomalies that require expert analysis and decision-making.

You should also consider the profile of the manufacturing workforce of the future and, in particular, how your business will require a significant number of multidisciplined professionals with working knowledge of critical disciplines such as business management, IT and engineering. They will be operating across multidisciplinary silos, connecting process and factories together, retrieving and interpreting data, and even protecting processes, products, business intellectual assets, supply chains and customers from security threats and intrusions.

There are many manual steps required before you can create a virtual model and link it to a physically observable product and process. For example, 3D printing depends on the links between product design, simulation, 3D printing software, 3D printing hardware and the actual 3D printing process. The steps might also include the capture of critical customer requirements and product concept designs, as well as simulations of the virtual product in collaboration with partners and customers before anything is actually configured or supply chain specifications produced or revised. Creating 3D simulations and proof-of-concept models can save a lot of time and money by ensuring the product is exactly as it should be before the manufacturing processes get underway. Another benefit from virtualization is that actual manufacturing processes can be simulated to make sure that they will run safely and efficiently, removing any doubt or trial and error from the process.

Overall, Industry 4.0 can improve how we all perform economically, socially and environmentally. However, in order to harness this potential, the challenge is develop a better understanding of human complexity and better methods to ensure that new technologies are trustworthy and adoptable by embedding human and societal values throughout their lifecycle.

By aligning our technology roadmaps with similar roadmaps for skills and capabilities, we can build the cross-disciplinary teams required to realize the benefits of Industry 4.0. Businesses will need to combine the trans-disciplinary knowledge of computer scientists, designers and industrial engineers with those of psychologists and social scientists to design and implement technologies for humans, rather than making them an afterthought.

These questions are the perfect prompt for tackling one of the most long-standing handicaps for UK manufacturers: the role of people in the context of productivity. Having the right skills and capabilities is fundamental to Industry 4.0 transformations. Organizations who are capable of identifying and addressing the gaps and complementing their existing company domain knowledge are likely to have a competitive advantage derived from their ability to utilize new technologies faster and in a more effective manner.

Necessary skills and capabilities can be built through a combination of internal training, the acquisition of new talent, collaborations with tech-solutions providers, and through research and academic institutions. However, capability building alone is not enough to capture and sustain the full value of Industry 4.0; it is imperative that these new capabilities become institutionalized and do not exist as separate or add-on skills. Companies will need to cultivate a culture that sustains new capabilities by integrating them with current capabilities and understanding them as essential to value creation.

That is where partnerships can really help. Working with and leveraging the assets and capabilities available in other sectors, and seeking the support of research organizations and universities can be a great start.

Customers

In the era of Industry 4.0, the customers' experience is everything. Whether that be through product evaluation or simply the experience of using the product, the goal is for the customer to gain value and for the manufacturer to build a good reputation. However, customer experience is difficult to evaluate. Yes, we can have surveys and ask the questions that relate to a product. However, we have seen in many examples this approach does not always work.

This is why big data and in-depth analytics is so important. It is no longer about a chief executive or their team determining the future of a product. It is about deep-mining research and product insights from embedded sensors that provide real-life information on how the customer uses the product, which when combined with sales information in true figures and customer experience.

Connectivity of products, manufacturing, supply chains and the use of big data analysis are fundamental to a better understanding of every stage of the product or service lifecycle. It can produce information that is individual, process or business specific, and knowledge that is vital to the product's design and a better customer experience.

Customer experience relates to the individual's acceptance that the product or service they bought offers good value for money. The acceptance is based on the quality of the product, the fitness for use and the cost in relation to competitive products. However, customer experience also relates to the way the customer relates to the goods, their relation to the direct channels for acquisition of the products and how they feel about the relationship between the parties involved in the production of the given product or service.

Say you buy a smartphone that costs £500, and then you discover that every application that you need has to be bought through the manufacturer. Would you be happy? The result would be a poor customer experience. So how do we manufacture products that our customers want? Through smart manufacturing and clever design; hence the need for smart factories and connected value chains.

5. UNLOCKING INDUSTRY 4.0

A recent study from Accenture[1] found that 86 percent of companies that began investing in technology actually failed to grow above the industry average, let alone see a return on investment. This is not surprising as the research conducted by the World Economic Forum[2] cited that despite the estimated investment of $1.2 trillion on digital transformation efforts in 2019, only 13 percent of leaders say their organizations are ready for the digital age. This is a particularly concerning set of statistics at a time when the pandemic is likely to prompt many financially vulnerable SMEs to explore investing in digital technology, anticipating that they will see quick results.

So, why are so many businesses, including SMEs, getting it wrong when it comes to adopting digital technology? Most businesses begin their digital journey by thinking about technology first, usually because they have been influenced either by the hype or the number of new suppliers pushing technology into manufacturing or even they may have seen a particular technology working well in another business setting. Unfortunately, one firm's digital best practice will not always apply to others. As a result, companies that begin with a vague strategy at best, which fails to consider the internal and external circumstances, will be taking a gamble which is likely to have minimal impact on their bottom line.

Rather than taking a technology-driven approach, firms that succeed in investing in Industry 4.0 and capturing above average returns on investment have adopted a value-driven approach which starts with thinking about how to enhance their value proposition in the context of their existing business model, before looking at suitable pieces of Industry 4.0 technology that can help achieve that goal. By taking this tack, some businesses may find that integrating digital technology may even pose a risk to their company. In the manufacturing case studies in Part 2 of this book, we will discuss how the deployment of digital technology in the wrong setting may contribute to companies weakening their competitive positioning by sharing trade secrets that formed part of their unique selling point to their customers. Industry 4.0 and related technologies should be used to enhance a company's existing unique selling points, rather than investing in parts of a business that have no bearing on attracting or improving the value and experience of customers.

Nevertheless, a value-driven approach to Industry 4.0 is easier said than done, because for a business to understand what the potential of digital technology is, business leaders need to understand the impact it will have within their wider value chain. If, for example, a Tier 1 manufacturer invests in technology, how will it affect their relationship with the original equipment manufacturer who buys their products? and to what extent can they leverage their position within their own marketplace and supply chain?

When it comes to Industry 4.0, inter-company collaboration is key. The relationship with partners across industry and with academia can play a powerful role in companies maximizing their use of digital technologies, de-risking investments in incremental stages, and developing the required skills and capabilities.

An example of success in this approach is a recent collaboration funded by Innovate UK. A Tier 1 manufacturer, its partners and customers worked together to convert existing vehicles from petrol to electric, replacing a traditional powertrain with an electric version driven by a large battery pack. The partners involved in the consortium worked together to create a digital strategy to maximize value for all parties involved.

As a result, an Industry 4.0 architecture was jointly designed and set up. A newly created factory was connected to its supply chain, enabling the partners to reduce time to market, improve product and process design virtually, improve product and process traceability, improve confidence in the business case for investment and, finally, reduce the overall risk of launching a product with the novel propulsion technology.

The project partners went back to the value-driven approach of analysing their business model and unique selling points, so they could make them relevant for the new digital systems deployed across their engineering and production processes. For example, analysis that came out of the value-driven approach included observations about the type of future customers, likely volumes and complexity of future products. A strategy for digital systems was created to maximize agility in response to OEMS' product and manufacturing design changes. As a result of this cohesive effort to deploy digital tools, a new design can be turned around within twelve months and highly flexible manufacturing facilities can cope with multiple product designs and varying levels of automation.

This collaborative approach is at the core of Industry 4.0. When compared with the previous three industrial revolutions, we see a distinction in terms of pace of change. The gap between the first and second being just over 100 years, second and third being just over 50 years, and third and fourth being around 30 years. We also see a distinction in the level of complexity that comes with the merging of physical, digital and virtual worlds which are enabled by data collection and connectivity across whole value chains. If we make an analogy with the driving forces behind each of the industrial revolutions, we can argue that the first was triggered by steam power technology, the second by electric power technology, the third by computing and automation technologies, and the fourth by connectivity and big data as a conduit to all other enabling technologies.

It is argued that Industry 4.0 is part of a data-driven future where intangible assets, such as data, are going to play a critical future role for manufacturers, particularly within automotive, when it comes to improving products and service offerings. This is evident in the WEF report, *New Paradigm for Business of Data Report*[3], which emphasizes that intangible assets (including data) make up to 84 percent of a company value. Another study also shows that in April 2020[4], the Chinese authorities began to recognize data as a new factor of production equivalent to and potentially replacing GDP in the future.

Understanding new business models within manufacturing and the many new revenue streams that can be created by collecting and integrating data across the lifecycle of a product is a collaborative task. It requires the involvement of internal and external resources in order to identify and evaluate the business case for investment. For example, with electric vehicle batteries, understanding the data linked to the life of a battery is critical to designing and manufacturing longer life batteries, as well as being able to re-use batteries packs for disassembly for a second life on other products. Understanding this opportunity and aligning the right resources and technologies at an early stage of product and process design can make all the difference between recovering value of investment and making a loss.

The lack of collaboration across the value chain is evident when we look at small businesses which constitute between 40–60 percent of the GDP in most countries, yet according to a recent report only 20 percent of European SMEs are investing in digitalization. This means that where perhaps larger companies up the value chain are investing in Industry 4.0 adoption, they will be doing so in a centralized manner and will therefore fail to reap the aggregated benefits across their entire chains. It is imperative that innovative partnerships between SMEs, large companies, tech companies, universities and governments de-risk and accelerate digital transformation for SMEs to achieve the full potential of Industry 4.0.

While technological development and change is continuing to increase at pace, manufacturing businesses need to resist the urge to rush into investing. Instead, they should take a rational and considered view of what is going to have the most impact for their own business. Early adoption of digital technology does not always have to be about immediate implementation and short-term results, but more about starting a diligent, incremental process that will yield results for years to come.

Innovating with business and operating models

Innovation strategy must be value driven, taking into account business models and operating models. Although these two concepts are tightly interconnected and mutually reinforcing, they embody different concepts, so it is useful to make a distinction between them.

Operating models are concerned with how an organization creates value. By operating model, we mean the value (in a product or service) that an organization creates, the locations where value adding work is done, the information systems that support operations, the supplier network and the management systems that coordinate the overall value chain. On the other hand, a business model is concerned with how an organization captures and delivers value. It is the way an organization realizes revenues (and profit or loss for profit-oriented businesses) by capturing and delivering value to customers. Understanding this distinction and designing an Industry 4.0 strategy that is aligned to support value at both levels is critical to success. In a pragmatic spirit, the focus in this book is mainly on two precepts:

Industry 4.0 adoption must focus on value drivers

Business should think value backward instead of technology forward. Focus on the value drivers in your industry or business in the context of your operational and business models. Then develop a compelling vision to inspire the organization and prepare internal skills and capabilities to work with partners, academia and government-supported pilots.

Industry 4.0 adoption must mobilize the entire organization

In order to be effective Industry 4.0 transformation must be driven from the top with clear business ownership aligning business strategy with business models, operational models and IP strategies. Companies focused on locking in bottom-line benefits quickly to prove value in the real world, whilst at the same time building capabilities and cultivating a highly flexible culture and infrastructure, will be positioned to capture more value from Industry 4.0 than its competitors.

Three common failure points

Finally, we will finish with three factors that regularly contribute to failure to capture value from Industry 4.0.

Lack of an Industry 4.0 innovation strategy

Often companies base their Industry 4.0 approach on the pre-condition that new technologies and innovation for the sake of innovation is good in itself. However, in order to capture value from Industry 4.0 your business will need to move beyond this view into a comprehensive strategy detailing what Industry 4.0 and digital innovation really means to your business across operational and business domains establishing clearly how investment in Industry 4.0 adds value to you.

Lack of an innovation system for Industry 4.0

Another common occurrence is the lack of resources and capabilities to deliver the Industry 4.0 strategy and to use digital technologies to innovate effectively. Most companies hire consultants, system integrators or other technology suppliers to come and implement best practices that work for other companies. This rarely works, as the best practices for other companies are not necessarily the best practices for your specific business and business model. Industry 4.0 adoption has to be methodical, pragmatic and tailored to satisfy the particular conditions and aspirations of your business. It is fine to seek resources and capabilities to support your Industry 4.0 journey, however, no partner will know your business as well as you and your team, and you should not outsource critical decision-making to a third party who does not have an interest in the long-term return on Industry 4.0 investment and sustainable competitive advantage.

Lack of innovation culture

Finally, companies seeking to innovate often forget this last factor. You can build the best strategy and develop the best innovation system in the word, but if your business does not have the culture to innovate and adopt Industry 4.0 technology at the right pace you will not be able to ever achieve a return on your investment. Those of us who have been involved in the traditional lean, lean sigma and six sigma methodologies, especially those emanating from Japan will understand the importance and emphasis placed in building the right culture, communicating the strategy and the importance of innovation right across the business.

Mastering how value is created is essential to the long-term success of any company. A business model expresses how a company creates, delivers and captures value. Being able to recognize where value is being created, distributed and captured and where it is not, therefore, plays an important part in designing an Industry 4.0 strategy that enables value capture. Getting the right skills and capabilities in place is a key success factor for Industry 4.0 adoption. You will need to identify your organization's gaps and complement the existing company domain knowledge in order to close these gaps in the areas of, for example, advanced analytics, artificial intelligence and IoT stack architecture.

These necessary skills and capabilities can be built through a combination of internal training, the acquisition of new talent, and collaborations with tech-solutions providers, research bodies and academic institutions. However, capability building alone is not enough to capture and sustain the full value of Industry 4.0. It is imperative that these new capabilities become institutionalized and do not exist as separate skills or add-on teams. Your business will need to cultivate a culture that sustains these new capabilities by integrating them with current capabilities and understanding them as essential to value creation in the age of Industry 4.0.

Notes

1. See the report at https://www.accenture.com/_acnmedia/thought-leadership-assets/pdf/accenture-unlocking-innovation-investment-value.pdf.
2. See http://www3.weforum.org/docs/WEF_Digital_Transformation_Powering_the_Great_Reset_2020.pdf.
3. Available at https://www.weforum.org/reports/new-paradigm-for-business-of-data .
4. See the article at https://www.chinadaily.com.cn/a/202004/10/WS5e903fd7a3105d50a3d15620.html.

PART 2

THE DYNAMICS OF VALUE FROM INDUSTRY 4.0 INNOVATION

6. INNOVATION FOR DIGITAL MANUFACTURERS

The speed at which global competition now moves demands that companies constantly find ways to innovate. Increasingly complex products, speed of innovation from new market entrants, as well as instant customer feedback via social media, are some of the frequently cited factors to justify why businesses have to create ways to collaborate and innovate to be competitive. Manufacturers widely recognize open innovation as a way forward, yet most continue to view it as a series of one-to-one collaborations. In Industry 4.0, the value will flow to those who can innovate in a network and who can use their IP effectively, so innovation relationships can be as tight or as loose as required to achieve a particular business objective.

Open innovation requires a constant flow of knowledge (inwards and outwards) regarding strategic positions, approaches, decisions, models and choices. One of the first questions to ask in this environment is: why should we be open or closed? Given the level of connectivity involved in the implementation of Industry 4.0 already discussed and regardless of your current position with innovation models, it is likely that your business will become more open.

As such, perhaps the more relevant questions will be: how open should your business be? in which relationships and with which partners should you be open? and in which areas or technologies should you be open or closed? This chapter will explore these questions.

The challenges of Industry 4.0 digital transformation

Innovation-driven transformation is a multidisciplinary effort that involves a combination of ideas, new technologies, a process of change, and the learning and development of new capabilities. Innovation is traditionally split into different types, most commonly referred to as incremental, radical, disruptive and open.

For the purposes of this discussion, we will use the working definition of innovation as: 'a new way to convert an idea into a commercial proposition that adds value and enhances an earlier solution or business model'. In the context of Industry 4.0, such innovation can materialize in the form of new or enhanced products, processes, services, technologies and business models.

To improve your chances of success in profiting from innovation based on Industry 4.0, your business should be clear from the outset about how the innovation is going to contribute to your current and future offerings and, in particular, how it will aggregate value to your customers. To achieve this level of clarity and purpose you need a good understanding of the dynamics of other businesses in your value chain: what are their key interests, capabilities, strategic intents and future targets? and how do they create value?

The objective here is to form a view of your business landscape, which in turn will support the formulation of a future business model. To facilitate this activity, you and your business will require an appreciation of processes, mechanisms and systems forming part of a networked innovation system which will emerge as a result of the Industry 4.0 connected value chains and will certainly include a great deal of inter-organizational relationships with the shared purpose of innovation and value creation.

The innovation process in connected value chains will follow the traditional open funnel with fuzzy ideas at the front end and clear commercial opportunities emerging from the innovation at the back end. Nevertheless, it is important to remember that innovation is not a linear process: it can be initiated from opportunities emerging at any point across the connected value chain. It is also a cumulative process with each innovation building upon previous knowledge and ideas. These outputs will disseminate widely among those taking part.

This is where we have to discuss the management of intellectual property and the shortcomings of traditional approaches to formal protection methods which normally do not offer sufficient protection and clarity in this networked innovation model. These traditional methods of IP protection are focused on giving protection to the rights of individual businesses, and fail to recognize and enable the new set of methods and models for knowledge sharing and protection in the context of Industry 4.0.

New IP protection models have to consider that network innovation is inherently uncertain. Value creation and capture in connected, digitalized manufacturing will depend on your ability to turn this uncertainty and IP risk into opportunity for new pathways to value. The businesses in the networked innovation model who appreciate the dynamics of inter-organizational innovation will have an advantage as they are likely to be aware of how others' choices and decisions influence the overall capture of value, which, in turn, will enable the to strengthen their competitive advantage and profit from the innovation.

Industry 4.0 is inherently an exercise in open innovation

Open innovation as a term is still new and often confusing to many businesses. There is a general misunderstanding between open and free innovation. In practice, open innovation is a process where a business uses external knowledge sources instead of free technology or funding from an external source.

There is no one model for open-source innovation that fits all. In the context of Industry 4.0, each business will have its own offer and position. As such, it should decide why, how and when it would be sensible to utilize open innovation models and the relevant degree of openness.

Businesses have alternative models to innovate and can traditionally select different approaches to suit each situation. However, in the current complex technological landscape, businesses are rarely able innovate in isolation without collaboration and input from partners and use cases.

This almost rules out the closed innovation models where the innovation process is internally focused within the boundaries of a single business. The closed innovation model normally demands close relationships with known and trusted partners in the value chain. This innovation approach traditionally is based on deep and continuous learning between partners and may lead to long-term dependencies.

On the other hand, open innovation is based on dynamic relationships beyond business boundaries, which are leveraged to increase your own innovation potential. Your employees, customers, suppliers, users, universities, competitors or companies from other industries can be integrated into the process. As a result, your business is more likely to create radical innovations. Nevertheless, uncertainty and risks are substantially higher. It is argued that these open innovation relationships are similar to the Industry 4.0 relationships required to roll out digital technologies across entire manufacturing value chains.

Depending on the model selected, innovation could emphasize either vertical or horizontal relationships involving customers, suppliers and even competing businesses. All of these could provide relevant sources of knowledge and information required to innovate. For example, customers can provide a first-hand account of their evolving requirements, then suppliers and competitors could provide ideas and insights that let you develop an offer for an unmet demand.

In addition to direct supply in the form of products or resources, there is a range of intangible knowledge, such as reputation, network connections and experiences, which is important to business development. In horizontal relationships, consultants, universities, research centres, other companies, innovation intermediators and funding agencies act as sources of knowledge for developments, such as information, technology and funding.

In these types of collaborative innovation models, the similarity between the knowledge bases, as well as value-chain positioning, has a significant influence on the possibilities and willingness to share or transfer knowledge and the opportunity for value appropriation.

The focus of customer-supplier relationships is typically on the supply of existing products, which thinking leads to incremental innovation. In customer-supplier collaborations, the vertical relationship often guides the businesses to behave in a one-way mindset where the customer is in control of the relationship and the supplier has limited choices and negotiating power. In these types of relationship, the feedback loops and two-way interactions are limited or even non-existent. In the case of SMEs in highly competitive manufacturing value chains, the opportunity to protect their innovation in such relationships is low as will be discussed in Chapters 9 and 10.

This was pointed out by one of the interviewees in our case studies representing a small manufacturing company based in the UK automotive manufacturing supply chain:

> Large customers want to own all the IP associated with the automotive products and processes related to their particular supply contract — they don't even consider allowing us to utilize the technologies in other industrial sectors.

In contrast, horizontal value-chain collaborations and alliances may promote equality between the actors. For example, one interviewee from another small manufacturing supplier described how the industrial collaboration with a supplier in the automotive industry is designed for mutual benefit:

When the innovation cannot be used for direct business benefit, we tend to compromise and look at options to benefit indirectly via supporting activities, such as using project outputs to improve our marketing, gain technology know-how, improve our brand perception and our team's skills.

Traditional models of innovation emphasize one direction of collaborations and often ignore the importance of understanding the networked innovation system itself, ie, how interactions, relationships and processes between businesses should be supported. Such innovation is paramount in the context of Industry 4.0's connected value chains.

The characteristics of Industry 4.0 connected innovation

In a connected value chain, businesses have a set of direct relationships when innovating. They also have a set of indirect relationships via its partners' networks of customers and suppliers. All these relationships develop and co-evolve in connection with each other.

The motivation and commitment to participate in connected value chains is different for each business, depending on its own targets, position in the value chain and IP strategy. These individual drivers inform the selection of innovation models (open, closed or somewhere in between) and the configuration of the network. Thus, the alignment of interests and interdependencies have an influence on the ongoing negotiations and contracts between partners in the value chain.

The type of relationship also has an influence on how innovation in a value chain performs. There is an intense debate in academia and practice as to whether tighter or looser relationships are more beneficial in the short and long terms.

Co-operation and collaboration within connected value chains generate trust and facilitate learning and the exchange of tacit knowledge between businesses. On the other hand, more open, loosely coupled networks with many weak ties and structural holes may have more advantages due to the fact that businesses can build relationships with multiple unconnected partners and explore opportunities.

Regardless of your preference, the implementation of Industry 4.0 in your value chain will require your business to participate in increasingly complex, integrated and yet loosely coupled networks. It is important to consider that these relationships will consist not only of formal business relationships, but also a number of informal relationships between businesses, workers and social networks, which may be an even more efficient way of finding new solutions.

Therefore, in order to maximize these formal and informal

relationships, your business should strategically consider the whole value chain, the positions of each actor and the relationships between them. If informal social interconnections of employees are identified as an important source of new knowledge, the IP strategy of a company should clarify to its employees how and with whom these connections could and should be configured. Still, according to our case studies, most innovation relationships are still based on contractual relationships, although more than two actors are involved in the innovation process. These multilateral relationships and contracts are typically utilized in collaborative research projects.

Managing collaborative innovation

The management of collaborative innovation, particularly in connected value chains represents a challenge for many manufacturers. It emanates from the complex dynamics of networked innovation: objectives, roles and relationships are likely to change during the different phases of innovation, depending on the particular technologies.

Your business, knowledge, IP and areas of interest will directly influence the role you play within the innovation model and how projects are configured. Capabilities and attributes, such as reputation in collaborations and technology know-how in a particular field, can provide advantages and attractiveness as a partner.

Clearly, the ability and willingness to collaborate will impact how you integrate internal and external interests at several levels of your organization. The role you play in this innovation process will certainly evolve and may also require the reshaping of the innovation network as its transitions from one phase to another.

The evolving nature of the value chain as an innovation network, its dynamics and timing pose challenges to management: for example, partners can be at different phases of evolution in how they innovate. For example, a particular partner might consider the objective of the innovation to be the exploration of new knowledge and future long-term business opportunities, while another may be focused on the operation of present business models and expect that the benefits will be realized in the form of short-term opportunities. This is certainly the case in value-chain collaborations involving research organizations, engineering consultancy firms and smaller manufacturing businesses.

The following scenario describes the dynamics of the process between two businesses, a product designer and a manufacturer (which was a customer). Their collaboration involved the development of a product (an automotive engine component) and the project took place over several months. As well as the product development, production processes were re-developed and optimized in the course of the project. Upon conclusion of the project, as agreed, the manufacturer commercialized the product.

The manufacturing business in this case focused on the development with an emphasis on the offering phase, as it had already identified a customer pull for the product within its existing value chain. Thus, the design company, which was involved in the development process, became used to a tighter relationship than other customers and thereby formed some misconceptions about how the collaboration would continue, eg, it expected that its employees would also participate in the implementation as deeply as in the development. As stated by one of the senior engineers in the product design business:

> Close to the end of the project, there was clearly a move to reduce the scope of the ongoing service support we anticipated from the customer.

Despite this minor friction post-project, both businesses described the collaborative innovation project as mutually beneficial. The representative of the design business mentioned that:

> Although we had the engineering and design know-how for the product development, we did not have the required capabilities and understanding of the actual OEM needs. Our team has shared knowledge and expertise with the manufacturing company and helped them build their capabilities to provide a full solution.

This project also demonstrates how the collaboration utilized intangible assets such as product IP, know-how and other tacit knowledge about the market to enable the development and commercialization of a new product developed by the two partners.

In the case of this project the relationship between the partners was clearly defined by the collaboration agreement, which set out the rights related to IP resulting from the project. However, the agreement did not address the IP issues related to background information. Nevertheless, both businesses recognized the tacit knowledge created in the project was valuable to their competitive advantage in the future and they would not have been able to create it by themselves.

In connected value chains, innovation is subject to individual business decisions driven by each of the partner's own operating models and decision-making processes. These decisions and operations of sub-systems cause emergent changes to the business environment and to the innovation networks.

This point was recognized by a senior manager we interviewed with expertise in industrial innovation at a research organization when he described the importance of understanding the positioning of partners in collaborative innovation:

> The success of collaborative and open innovation projects in the manufacturing environment depends on the positioning and long-term strategic fit of each of the partners. Our team always consider the various scenarios in relation to our partners and their strategies upfront before entering into the projects. It is difficult to do this if you don't know the partners you are working with or their strategic intentions.

Understanding the core capabilities, expertise, experiences and track record of potential partners also has an influence innovation on activities within the connected value chain. Even tacit knowledge at an individual level has an important influence on the capability of you and your partner to utilize innovation outputs.

Capabilities related to collaboration, team working and communications will be critical to collaborating effectively in the complex environment created by the connected value chains. Such capabilities support the creation and maintenance of a mutually beneficial innovation model, contributing to the success of the entire value chain.

Configuring the value

At the relationship and collaboration level, the management of networked innovation is all about balancing different interests and engagement models. In other words, at the value-chain level of innovation co-ordination is based on both control governance and self-organization. The balance between control governance and self-organization is different within distinct innovation models. Similarly, to the innovation process itself, the businesses involved in innovation are dynamic and evolving. In the 1990s, Ring and Van de Ven[1] proposed a cyclical model of network formation, encompassing four distinctive activities taking place amongst network partners as:

- negotiations
- commitments
- executions
- assessments

Each of these plays a generative role in the development of further activity in a value chain. The contribution of their perspective across the in the processes is particularly valuable to deepening the understanding about the balance between control and self-organization within the management of innovation. It highlights the ways in which value-chain partners adapt and re-evaluate their roles and commitment, as a response to decisions and acts of each other.

You will only be able to interpret the choices and decisions of each business in the connected value chain when you understand the strategic aims and the agendas of your partners. During all steps toward networked innovation, your business has to deal with explicit and tacit knowledge needs, the search for competencies, and the use of available IP.

It is critical to understand the targets and motivation of each partner involved. Businesses in the connected value chain should be aware of changes, emerging threats and opportunities within the digitally connected value chains. In an effort to improve your understanding of potential value-chain partners you should ask yourself questions such as:

- Why is this partner open in the innovation process?
- How open is the partner?
- Why should our business be open to the partner?
- How open should our business be?
- What are the other businesses' interests and motivations to collaborate and give their best efforts?

It is important that your business also have clarity on roles, responsibilities and potential conflicts of interest in this connected environment. In doing so, you should consider questions such as:

- On which areas and technologies are we open or closed?
- What competencies and capabilities are unique to our business?
- To which partners are we/should we be open?

Based on negotiations and contracts, you will be able to configure an appropriate form of networked innovation and commit to it. This is critical to anticipate potential issues and to create relationships in a mutually beneficial manner. This could be done through answering the questions again from the viewpoints of other partners in the network:

- Who are they are willing to collaborate with?
- Which issues and knowledge are they prepared to share?

By taking these strategic and transparent steps to configure the network you will be able to recombine knowledge to build unique innovation and aggregate value across implementation and execution of new solutions.

Within Industry 4.0 digitally connected value chains, you would benefit from considering in advance what your exit strategy is and any potential issues related to interoperability. Businesses can potentially end up in situations where they have invested heavily in developing and building a particular innovative process, product or service, but fail to plan appropriately for how the business will capture value from the innovation.

In other words, it is so easy to get carried away with the exciting technologies and innovations resulting from the adoption of Industry 4.0 that you may have not thought over the possible changes in the value chains, the possible influences on the competitor landscape and the demand for your products or services their products. Exit strategies are often forgotten and only considered when relationships have already broken down, which limits options for a pro-active mutually beneficial exit pathways.

As such, your business should consider the scenarios and pathways for development of digitalization across the value chain: identify partners beforehand in order to form an exit strategy and consider the potential post-contractual issues.

It is unlikely that a single business will control and manage the entire connected value-chain innovation system. On the contrary, every business will have to manage its own position and role. This will require systematic and strategic analyses of the innovation environment, the partners in the value chain and their objectives in the short and long term.

Based on your analyses, you will be able to make conscious decisions, take calculated risks and leap at potential opportunities. These strategic analyses of the innovation and value-chain environment require the right skills and resources to foster an open dialogue with innovation partners, especially with new partners with whom it might be hard to find a common ground.

Nevertheless, when considering the context of the connected value chain, you can easily understand how important it is to go through other businesses' value offerings regularly and hold strategic discussions with other partners in order to keep in touch with forthcoming changes. There again, you also know that in value chains some of the businesses come from different cultures and have different approaches to creating, delivering and capturing value.

To summarize this chapter, every business is able to manage its own position and role within the Industry 4.0 connected value chain and the emerging networked innovation system. However, this requires systematic and strategic analyses of the innovation environment, potential partners and their strategic objectives.

Notes

1. Ring P. S and Van De Ven A. H. (1994), 'Developmental processes of co-operative interorganizational relationships', *Academy of Management Review*, 19(1): 90-118, available at https://doi.org/10.2307/258836

7.
APPROPRIATING VALUE

There are a number of theories dealing with the creation and capture of value from intangible assets in disciplines such as law, economics, business management and many others, most of which have limited application from a practical point of view. On the other hand, the theories and models for intellectual property found in the academic and practitioner literature of business management are well aligned with Industry 4.0's value-chain scenarios and typically offer a view of value capture from an individual business point of view within a given value chain.

These theories often consider the rules of the system as a given and describe how businesses can or should behave within these boundaries. They are interesting in the context of Industry 4.0 as the mechanisms and models for protecting IP often consider the management and adoption of new technology and innovation within a business or a business relationship. This typically spans the entire process from an initial idea all the way to an innovation on the market. The tendency is to split the technology and innovation process down into two sub-processes. These are labelled creating value and capturing value.

The first sub-process, creating value, focuses on the generation of innovation (Teece, Pisano and Shuen, 1997[1]; Amit and Zott, 2001[2]; Thomke and von Hippel, 2002[3]). The second, capturing value, focuses on how companies can best exploit their technologies and innovations (Teece, 1998[4]; Chesbrough and Rosenbloom, 2002[5]).

Appropriability regimes theory

One IP theory of particular importance for our discussion in the context of Industry 4.0's connected value chains is about how value is generated and captured by businesses in a given set of relationships, the appropriability regimes theory. This theory offers a method to assess some of the risks associated with changes in relationships in the context of value generation and capture. In doing so, the theory can support your understanding of how Industry 4.0 affects the manufacturing value chain by changing its relationships through the adoption of Industry 4.0 technology, which in turn affects the existing innovation and competition paradigms.

Appropriability in a digital world

Appropriability is a critical factor for the effectiveness of innovation. If a company innovates well but fails to appropriate the innovation, its competitors will imitate and commercialize the innovation without the costs of research and development incurred by the original innovator. Without appropriation, the initial company has no incentive to invest in innovation. From this point of view, the legal, business and economics theories are directly linked because, at an individual level, a company must be able to appropriate its innovations in order to generate value, which is then re-invested in order to continue innovating.

Over three decades have passed since Teece's (1986) article 'Profiting from technological innovation: Implications for integration, collaboration, licensing and public policy'[6] conceptualized the theory of appropriability. Nevertheless, appropriation is still a popular topic. Recently, a number of fundamental questions regarding this area of research were explored and answered. Of particular relevance is the fact that the concept of generating value from innovation has been transformed over the last few decades from a concept based completely on financial returns or value from innovation to a model which expanded into other forms of intangible value, such as competitive advantage, brand recognition, time to market and many others.

It is an example of how companies have changed their IP strategies in order to effectively protect and appropriate value in a dynamic context where there is a change in paradigm. It is also becoming the case for manufacturers, where appropriability as a theory can be used to support the understanding and identification of actions to address risks and opportunities.

The appropriability regime framework

The term appropriability in this context is used to characterize the extent to which manufacturers are able to obtain a return equal to the value created by their innovations. Expanding on this definition, we will focus on the circumstances which enable the value generated by technological innovation to be captured.

Furthermore, the term 'appropriability regime' will be used in this context to understand a business's ability to capture the value generated by Industry 4.0 innovation in connected value chains for manufacturing industry in the UK as documented through four case studies.

This book relies on the author's own research that has built upon the current theory of appropriability in order to develop a method called the Manufacturing Appropriability Regime Construct (Marc). The data is consolidated in four case studies representing typical manufacturing value-chain relationships from transactional at one end of the scale to strategic relationships at the other. Through their experience, the aim is to give insights into manufacturing digitalization, collaborations and the impact of Industry 4.0 on manufacturers' appropriation of value in different relationships within the context of distinct phases of digitalization maturity

The current theory of appropriability regimes is composed of two parts: firstly, the nature of knowledge in the particular innovation; secondly, the strength of the legal protection of intellectual assets. The appropriability regime framework provides a perspective on the business's particular position regarding the prospects of its innovations being quickly replicated by the competition. The framework has two main divisions. The first of these is inherent replicability and the second is IP rights.

Each of these two broad divisions has two subcategories, inherent replicability is divided into easy or hard, referring to the level of difficulty involved in imitating the innovative product or service. On the other hand, IP rights are divided into loose or tight intangible asset protection, referring to the level of IP protection available in the context of a particular innovation.

If we were to plot the two categories and subcategories into two axes, we create a four-box grid. On the bottom left-hand side, we see the weak appropriability quadrant where an innovation is easy to replicate and IP rights are loose. At the top left-hand side and bottom right-hand side, we see two moderate appropriability quadrants where the innovation either is hard to replicate but the IP rights are loose or where the innovation is easy to replicate but the IP rights are tight. Finally, at the top right-hand side we have the strong appropriability quadrant where the innovation is hard to replicate and the IP rights are tight.

These four quadrants can be used to understand whether the innovator's position is weak, moderate or strong in regard to capturing value from innovation as demonstrated in the traditional appropriability regime model in Figure 1 below. Each of the fields will be briefly discussed in the following sections in order to explore the analysis of each actor in each of the case studies.

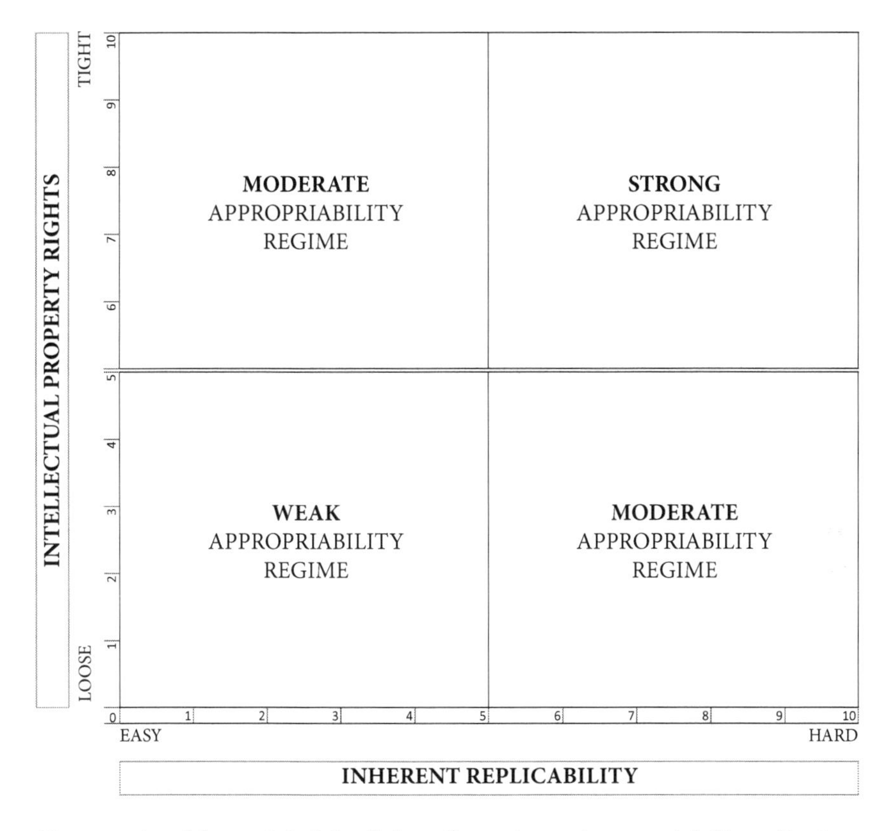

Figure 1: Marc Model (Manufacturing Appropriability Regime Construct)

The weak appropriability quadrant

Innovations generated by businesses located in the weak quadrant are easy to imitate and the IP rights available to protect the innovations are loose. Typically, if the businesses located in this quadrant do nothing to improve the IP protection for their innovations or develop disruptive innovation, they will fail to appropriate value from innovation.

Businesses that continue to focus only on the launch of new products, services or processes with small and incremental adjustments to the existing products or processes typically stay in this weak position and become a producer of standard or commoditized products or services in a homogeneous market.

The moderate appropriability quadrants

On the other hand, businesses positioned in the moderate quadrant in the bottom right-hand corner have a product or process that is hard to replicate, however, their IP rights are loose. Businesses in this quadrant have inherent organizational core competencies which are not protected by law but are difficult to replicate. It is generally more difficult to copy organizational processes and positions from other businesses, as copying and knowledge transfer are difficult between businesses with different competencies and systems.

Businesses in the second moderate quadrant in the top left-hand corner have easy-to-replicate products or processes, but their IP rights are tight. Businesses in this position are normally in industries where patents are strong and protected by law. Copyright and trade marks are strong because they are important for the customers.

The strong appropriability quadrant

Finally, businesses located in the strong quadrant in the top right-hand corner have hard-to-replicate products or processes and their IP rights are tight, so it is not only difficult to replicate their products or services, but it is typically illegal.

Businesses in this quadrant are normally highly innovative; they make use of both incremental and radical innovation. Theoretically, all companies should strive to get to the strongest possible position given their business models and markets. If they are in a strong position, they should strive to maintain the position.

The manufacturing appropriability regime construct

In order to examine the impact of Industry 4.0 on manufacturing appropriability, we have created a model based on the appropriability theory which was designed to improve its application to real-world situations. This model sub-divides each of the two areas of appropriability into three indicators which will be presented here as a model for identifying and evaluating appropriability indicators for the two parts (IP rights and inherent replicability) of the appropriability regime framework.

IP indicators

Practical means of protection

This indicator is designed to assess the practical steps taken to implement practical means of protection for innovation. As part of scoring this indicator in the business context, you should evaluate policies, procedures and cultural aspects of the businesses involved in the process of innovation. These include for example, whether the business has a clear policy defining what IP is important for the business and also whether employees know about and are trained to follow its procedures.

Contractual position

Contractual positions emerged as critical to the strength of IP rights during data collection, because they made data, information and even registered IP such as patents impossible to appropriate. This was the case particularly with contracts for the supply of products or services with clauses requiring the transfer of all current and future data and IP on the product to one of the contracting parties.

IP rights protection

This indicator is designed to assess the actual protection mechanisms utilized by a particular business in the context of the innovation to be evaluated. As part of this assessment you should consider the use traditional forms of registered IP, such as patents and trade marks, as well as unregistered forms, such as trade secrets and copyright, which may be more suitable for the protection of digital intangible assets shared among Industry 4.0's connected value chains.

Replicability indicators

The nature of knowledge

This indicator is designed to evaluate the nature of knowledge in the context of a particular innovation. It focuses on the extent to which the knowledge is explicit or tacit and also how consolidated or codified the knowledge is within a value chain. This is a critical indicator in the context of digitalization which normally codifies as much knowledge as possible in the process, product or business subject to the digitalization.

Technology readiness

Technology and manufacturing readiness levels (TRL and MRL) are well known in engineering as a method of indicating product or process maturity. The TRL and MRL levels scale starts at Level 1 where basic scientific research is performed to observe and document the concepts of a particular technology and finish at Level 10 where the technology is applied at scale in multiple sectors. The TRL and MRL levels are used in the context of appropriability regimes as an indicator of potential replicability. Higher and more widespread TRLs and MRLs result in increased chances of imitation by competitors.

Competency required

This last indicator is designed to evaluate the levels of skills and competency required to replicate a particular innovation. For example, if the innovation involves common competencies which are widespread across the manufacturing value chain, it is likely that competitors will have a higher potential to imitate. On the other hand, if the competencies required are new or scarce, it is less likely that competitors will be able to easily replicate the innovation.

All of the above indicators were designed with a 1 to 10 scale for the evaluation (where 1 represents low and 10 represents high). Furthermore, a scale of 1 to 10 for the assessment of the strength of the IP on the Y axis, which combines the three indicators for IP strength, and also a scale of 1 to 10 for the difficulty of replicability on the X axis are shown in the Marc Model (see Figure 1).

With the scales in place for each axis, the focus turns to the representation of each business in the case studies. This process began with an evaluation of each individual business in regard to each indicator on a scale of 1 to 10 in the context of each case study.

In the next step of the Marc analysis, we have a weighting system for the appropriability indicators in the context of each individual business model and value proposition. This weighting system is critical in order to effectively analyse the appropriability regime of a particular business as, for example, the value of a patent for a manufacturer will be different in cases where there is a contractual obligation to transfer this patent to a third party. This is common practice in automotive manufacturing for example, where the OEMs almost invariably own the IP on products and process made by the Tier 1 and Tier 2 manufacturers. Such practices may render strong IP protection through patents for example irrelevant in the context of value appropriation.

This is the case of manufacturers operating a make-to-print business model (limited or no product IP required as the blueprint for the product is provide by the customer) in highly competitive value chains such as the automotive manufacturing in the UK where value is only generated through the sales of commoditized products. To address these cases, the weighting was applied to the scores of a the particular business model between 1 and 5 (1 meaning the indicator is not important to the value proposition and 5 meaning the indicator is very important to the value proposition of the business in relation to the particular case study).

Once again, interview information regarding each of the participants' business models and each product, process and manufacturing technology, as well as the contractual information regarding each case study, was used in order to allocate a weighting to each indicator for each business in each of the case studies.

Finally, a formula was developed to multiply the indicators by individual weighting scores, adding the results of each indicator within a category and dividing the result by the number of inputs in order to arrive at the weighted scores. The resulting score was then normalized to the Marc framework scale.

In order to validate the scores and weights, once the calculation was completed, interviewees from each of the businesses involved in the case studies reviewed the scores and weights. The independent evaluation conducted in advance was then presented. Any differences and adjustments were disccused.

The results of the data analysis were then plotted onto the Marc Model utilizing distinct shapes to represent a particular business within a case study, as demonstrated in Figure 2.

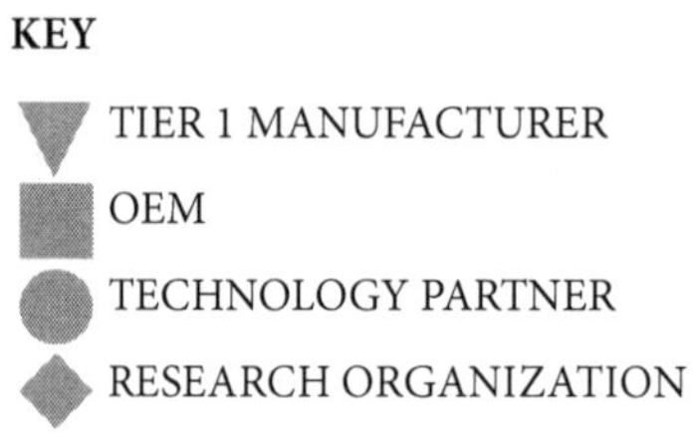

KEY

▼ TIER 1 MANUFACTURER
■ OEM
● TECHNOLOGY PARTNER
◆ RESEARCH ORGANIZATION

Figure 2: key for Marc Model case studies

The next chapter will discuss the cases study selection and the data collected. This will be followed by a chapter exploring each case study and a discussion of the findings from the Marc Model analysis.

Notes

1. Teece, D.J., Pisano, G., and Shuen, A. (1997) 'Dynamic capabilities and strategic management'. Strategic Management Journal, 18, 509-533. doi:10.1002/(SICI)1097-0266(199708)18:7<509::AID-SMJ882>3.0.CO;2-Z
2. Amit, R., Zott, C. (2001), 'Value creation in e-business'., *Strategic Management Journal,* 22(6), 493. doi:10.1002/smj.187.
3. Thomke, S., and von Hippel, E. (2002), 'Customers as innovators: a new way to create value', *Harvard Business Review,* 74-81.
4. Teece, D.J. (1998), 'Capturing value from knowledge assets: the new economy, markets for know-how, and intangible assets', *California Management Review,* 40 (3), 55–79.
5. Chesbrough, H., and Rosenbloom, R.S. (2002), 'The role of the business model in capturing value from innovation: evidence from Xerox Corporation's technology spin-off companies', *Industrial and Corporate Change* 11 (3), 529-555.
6. Teece, D.J. (1986), 'Profiting from technological innovation: implications for integration, collaboration, licensing and public-policy', *Research Policy,* 15, 285–305.

8.
PERFORMANCE IN THE VALUE CHAIN

To provide empirical evidence of the changes to appropriability regimes as a result of Industry 4.0's deployment across the value chain in manufacturing, this book draws on four case studies in which data was collected and aggregated to answer the how or why questions within a practical context. In the typical manufacturing value chain, there are many types of relationship between businesses, varying from the transactional at one end of the scale to strategic at the other. Multiple factors determine the nature of these relationships, such as duration, expectations, liabilities, level of interaction, goals, benefits and risks.

Transactional relationships are more prominent in the traditional vertical value chains where supplier and customer relationships are negotiated at arm's length. Strategic relationships are more associated with horizontal value chains where businesses seek to leverage each other's expertise for mutual benefit.

Manufacturers in the automotive value chain, for example, are involved in both types of relationship during the lifecycle of a product, service or technology. As an example, automotive product development, including the manufacture of components, is a complex process that involves many businesses across the value chain. These businesses normally start the development process with strategic collaboration projects where they work to create future products or processes. Once this phase of the lifecycle is completed, when products and processes are mature and production ready, the nature of the relationships tends to change towards a transactional basis, typically as supplier and customer.

The selected case studies discussed here aim to provide insight into manufacturing digitalization, collaboration, the impact of Industry 4.0 on automotive manufacturers and the appropriation of value in different relationships within the manufacturing value chain. The four case studies illustrate the range of relationships within the typical manufacturing value chain and the progress a manufacturing company like yours will take through various degrees of product and process digitalization.

These case studies represent collaborations, supplier-customer relationships and variations of projects within automotive manufacturing in the UK which typically fall under two categories (product or process development). The first type of project is focused on product development either for cost reduction, performance enhancement or both. The second type of project is normally focused on process / manufacturing technology development where the aim is operational improvement targeted at cost reduction.

In order to demonstrate the most common manufacturing relationships four key intersections of product and manufacturing technology were created as the basis for the four case studies.

- **Current product manufacturing with current manufacturing technologies**: projects where the nature of the relationships is transactional as there is no need for strategic relationships to develop new products or processes.

- **New product development with current manufacturing technologies**: projects where the relationships evolve from a strategic collaboration developing a new product a transactional relationship for the supply of this product to a customer (product R&D collaboration and supply contract).

- **New manufacturing technology development with current products**: projects where the relationships also evolve from a strategic collaboration, which in this particular case is focused on developing new manufacturing technologies, but which nevertheless result in a transactional relationship for the supply of a product to a customer (process R&D collaboration and supply contract).

- **New product development with new manufacturing technology development**: projects in this intersection are typically more complex and strategic as they involve a greater level of collaboration between the businesses in order to develop a new product and a new manufacturing technology which is interdependent, as the production of the new product depends on the new manufacturing technology. This type of relationship also evolves from a strategic to a more transactional relationship, however, the collaborators have a higher level of dependency at the transactional stage of the relationship (product and process R&D collaboration and supply contract).

With the above needs in mind, a set of case study selection criteria was created and deployed:

- The involvement of multiple actors in the automotive manufacturing value chain.
- The ability to demonstrate strategic and transactional relationships in projects for development of new products.
- The ability to demonstrate strategic and transactional relationships in projects for development of new processes or manufacturing technologies.
- Access to the data related to the businesses in the value chain. These included the interviews and the actual contractual agreements governing the relationships.

The next section will provide a summary of the data collected in each of the case studies.

Case data

The empirical data collected for the case studies included a set of 31 in-depth interviews and 11 contractual reviews.

The contractual reviews included the following types of agreements :

- Collaboration agreement.
- Supply contract.
- Non-disclosure agreement.
- Sample employment contract for Tier 1 employees.

Interviewees included those representing:

- Each of the parties subject to each of the case studies.
- Multiple administrative departments from each of the parties subject to each of the case studies including sales, commercial and legal departments.
- The engineering and R&D departments for each of the parties taking part in the case studies.
- Operations management.
- Senior management and directors at each of the businesses participating in the case studies .

The data collected via the interviews was utilized in combination with the contract analysis in order to understand the impact of Industry 4.0 in the context of each specific scenario. In addition, all of the interview data from across the case studies was used in an effort to identify common themes representing the main challenges across the scenarios and to shed light on the wider context of the connected value chains.

Impact of Industry 4.0 on appropriability

The following observations can be made about the findings revealed by the Marc Model analysis in regard to the impact of Industry 4.0 on the appropriability regimes of manufacturers in the context of the four case studies:

- In general, the implementation of Industry 4.0 alters the appropriability regime and leads to stronger positions for OEMs and weaker positions for Tier 1 manufacturers.
- Apart from the OEMs, as a result of utilizing Industry 4.0 technologies in the case studies, all businesses demonstrated a negative shift in inherent replicability, ie, it is easier for other businesses in the value chain to replicate your innovation.
- The Tier 1 manufacturer suffered the highest level of detriment to its appropriability regime in the case study where current products were manufactured utilizing new Industry 4.0 technologies (case study 4). It was a value chain with a high level of similar competencies and knowledge which resulted in innovations which were easier to replicate by other businesses.

These general observations are associated with a number of factors which are likely to influence the appropriability regimes and the IP strategies in Industry 4.0 connected value chains, which include:

- The formation of highly connected value chains with multiple businesses.
- The shift from tacit (non-documented) to codified (coded into a digital system) knowledge.
- The formation of value chains with a high intensity of data/ information exchange.
- The shift towards the digital engineering and digitalized product lifecycle management across the value chains.

- The use of out-of-date contractual agreements conflicting with current business strategies.
- The lack of IP strategy or the use of inadequate IP strategies and tools.
- Increasingly blurred industry boundaries where similar competencies can be leveraged to disrupt incumbent manufacturing businesses.

These brief observations emanating from the Marc Model provide a different perspective on the data collected though the interviews and the contractual analysis which exposes a unique view of the impact of Industry 4.0 adoption at different levels of the value chain.

This approach could be used to help your business to evaluate the impact of Industry 4.0 on its appropriability regimes for different scenarios. In addition, the evaluation can also be used to support decision-making when, for example, considering a new type of relationship or business models emerging from the implementation of Industry 4.0 and the changes that may be required to IP strategy. Developing an effective IP strategy is of paramount importance for manufacturers seeking to secure a strong return on investment especially in applications based on connected value-chain relationships.

In the next chapter each of the four case studies will be discussed in turn to provide a practical analysis of what exactly is changing in the way manufacturers appropriate value from innovation.

9.
FOUR CASES OF APPROPRIABILITY

This chapter explores four case studies of how value is now being appropriated in manufacturing, first in the current state of value chains and then for three progressively sophisticated levels of Industry 4.0:

- Digitalization of product development
- Digitalization of product development and manufacturing execution
- Digitalization of manufacturing execution for current products

Changes in the relative position of each business within these value chains are highlighted: for original equipment manufacturers, for Tier 1 and Tier 2 suppliers, for technology providers and for research institutions.

Current state

This case study gives a baseline, covering the current state of relationships in the automotive manufacturing value chain. It can be used to evaluate the impact of the change brought about by the implementation of Industry 4.0 and to monitor and assess the progress of manufacturing relationships and the appropriability position of each business before and after the adoption of Industry 4.0 taking place in each of the other three case studies.

Its data relates to a value-chain collaboration project aimed at further developing and optimizing an automotive component used in internal combustion engines. In addition to the product development and optimization, the project also included the re-commissioning of the manufacturing processes and operations required to make the component. In this particular case, the earlier iteration of the product was already in production in another manufacturing plant.

This project was part of a contract for the manufacture and supply of a product between a Tier 1 manufacturer and an original equipment manufacturer in the automotive industry. In this particular case, there were contractual relationships already in place between the OEM and the Tier 1 manufacturer and between the Tier 1 and the Tier 2 manufacturers which governed the ownership of intellectual property in relation to innovations in the context of the product being manufactured. The project took place over twelve months and involved five parties from across the automotive manufacturing value chain:

- A vehicle manufacturer, the OEM.
- A Tier 1 manufacturing business that supplies parts to the OEM (the leading party in the collaboration project).
- A technology provider who provides consultancy and engineering services in the automotive value chain.
- A research institution which provides applied research in advanced manufacturing.
- A Tier 2 manufacturer that supplies components to the Tier 1 supplier.

In addition to the contracts for supply and manufacture between the OEM and Tier 1 supplier and between the Tier 1 and Tier 2 suppliers, the other members of the project were also bound by a contract for research and development, establishing the roles and responsibilities of each of the parties, as well as the rules regarding IP.

This particular project provides a baseline representation of current technology in products and manufacturing processes with limited integration and data exchange across the value chain. The collaborating parties utilized tools and techniques that pre-date Industry 4.0's integration technologies and practices which are part of the other case studies. In this regard the individual businesses across the value chain operated in isolation with minimum data exchanged between the parties.

According to the data collected:

- The Tier 1 manufacturer is in the moderate quadrant, with a medium level of IP protection at 4.7 and a medium level of inherent replicability scoring 5.5.
- The OEM is positioned in the strong quadrant scoring 6.6 for IP and the same inherent replicability score as the Tier 1 manufacturer at 5.5.
- The technology partner is also positioned in the strong quadrant, albeit with a stronger score for both areas: IP at 8.4 and inherent replicability at 7.1.
- Finally, the university partner, scoring the highest possible score for IP, is also positioned in the strong quadrant with scores of 10 for IPR and 6.1 for inherent replicability.

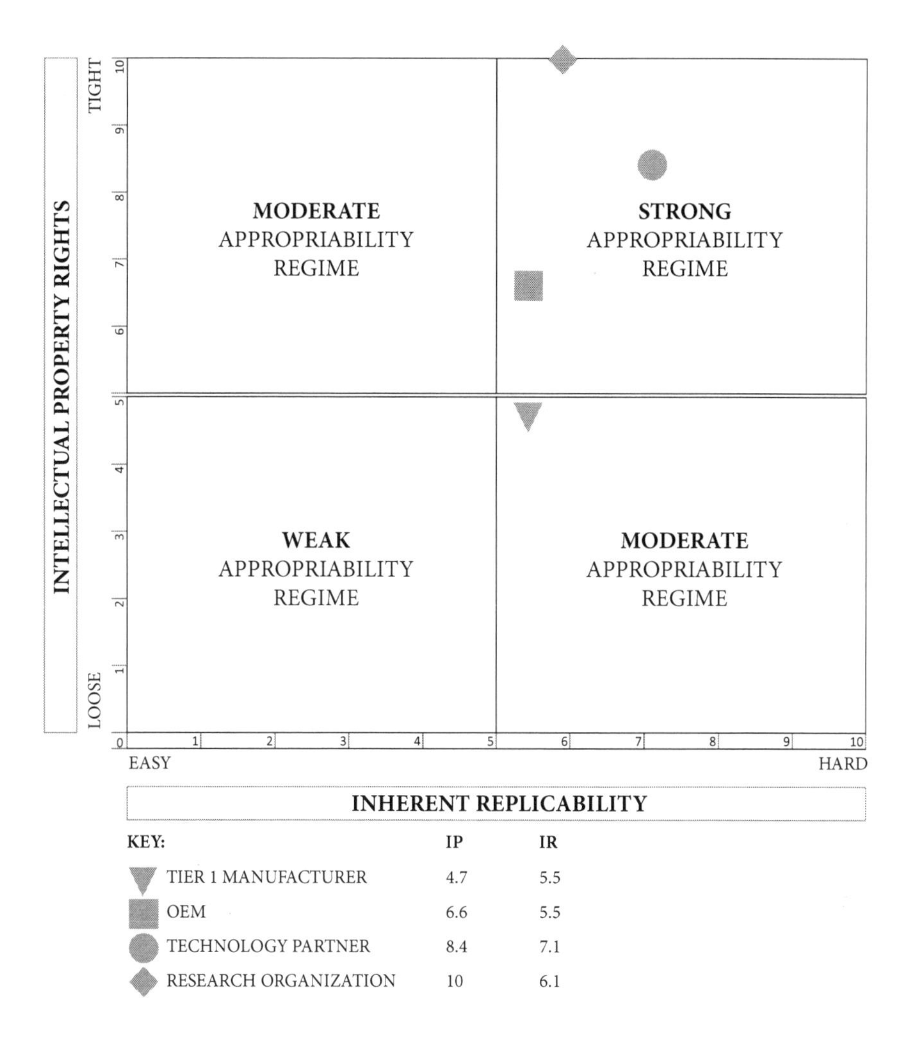

Figure 1: the position of each partner in Case Study 1 (current state) according to the Marc Model analysis

In summary, the analysis of the Case Study 1 indicates that the technology partner and the research organization were more likely to appropriate value from innovation created as part of this project. The OEM was close behind, still in the strong appropriability quadrant and the Tier 1 suppliers was the only business in the moderate quadrant.

Digitalization of product development

This case study provides an example of a collaborative relationship focused on the development of a product, followed by a contract for the manufacture and the supply of the new product utilizing current manufacturing technologies. The collaboration project took place over two years and involved five parties from across the automotive manufacturing value chain:

- A vehicle manufacturer, the OEM.
- A Tier 1 manufacturing business that supplies parts to the OEM (the leading party in the collaboration project).
- A technology provider who provides consultancy and engineering services in the automotive value chain.
- A research institution which provides applied research in advanced manufacturing.
- A Tier 2 manufacturer that supplies components to the Tier 1 supplier.

The aim of this project was to develop a new product automotive component capable of achieving higher functional performance compared to its predecessor. The main objective of this product improvement was to achieve lighter weight in order to reduce the overall emissions of carbon dioxide for vehicles utilizing this component and to reduce the overall product costs. The project was governed by a collaboration agreement which established the roles and responsibilities of each party, as well as the rules surrounding background and foreground IP.

In contrast to Case Study 1, this project utilized the latest digital technologies for product development, allowing the five collaborators to exchange product-related data and know-how to accelerate product development by running concurrent product design, design for manufacture and multiple product-related simulations. As such, this project provides a representative case study of the use of Industry

4.0 technologies for product-development relationships across the manufacturing value chain.

Upon the successful completion of this project, the Tier 1 manufacturer was awarded a contract for the supply of the product to the OEM for a period of five years. Subsequently, the Tier 2 supplier was awarded the contract for the supply of components to the Tier 1 supplier. Both contract wins were seen as success indicators in terms of both project outcomes and return on investment.

According to the data collected:

- The Tier 1 manufacturer is now in the weak quadrant with a lower level of IP protection at 4.3 compared to 4.7 in Case Study 1. Its position regarding the inherent replicability is also lower with a score of 4.4, compared to 5.5 in Case Study 1.
- On the other hand, the OEM position has improved and remained in the strong quadrant scoring 7.1, compared to 6.6 for IP and the score for inherent replicability has also increased to 6.1, compared to 5.5 in Case Study 1.
- In contrast, the position of the technology partner in this particular case has suffered. Whilst still positioned at the strong quadrant, the IPR scores slightly decreased to 8.3, compared to 8.4, and inherent replicability has decreased to 6.1 in comparison to 7.1 in Case Study 1.
- Finally, the university partner also remained in the strong quadrant, but also saw a decrease in both scores with IP at 9.7, compared to 10, and inherent replicability at 5.2, compared to 6.1 in Case Study 1.

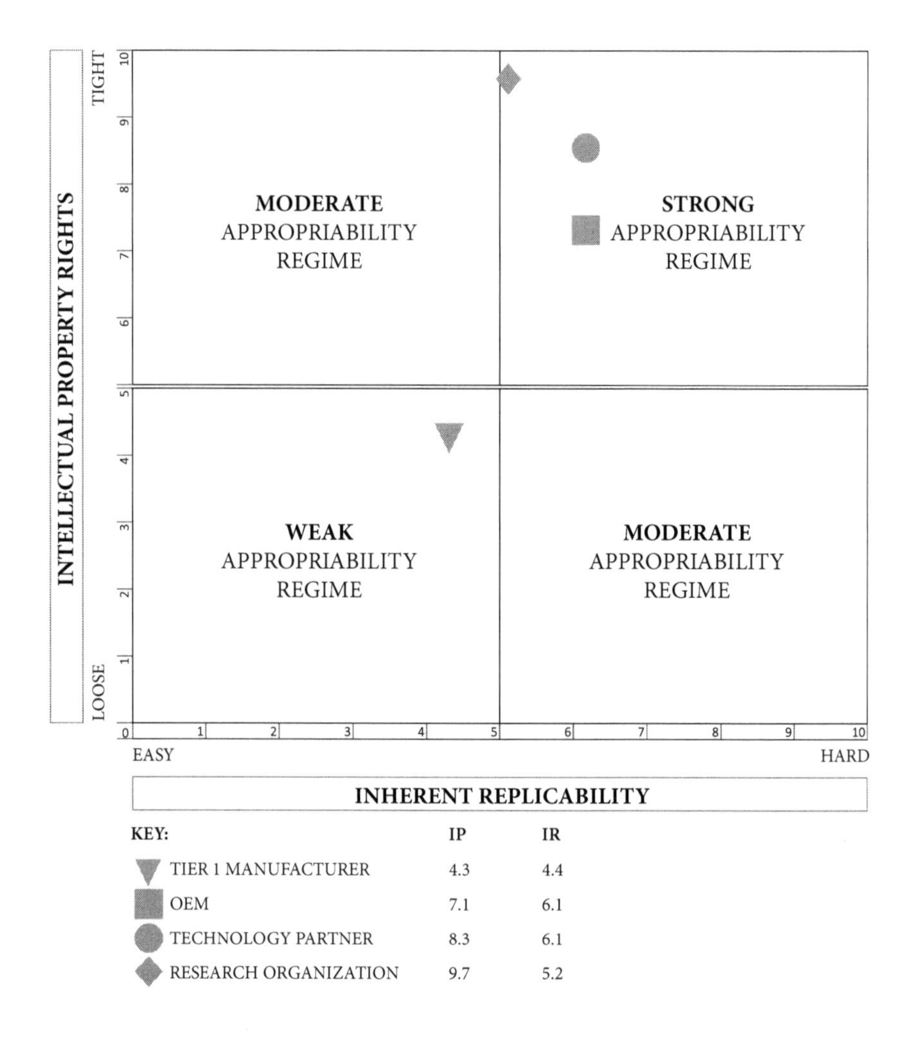

Figure 2: the position of each partner in Case Study 2 (digitalization of product development) according to the Marc Model analysis

In summary, the key change following the implementation of Industry 4.0 technology in this case has been the position of the Tier 1 manufacturer which has weakened, whilst the position of the OEM has gained strength. The main influences on this shift were:

- **Changes to the contractual position**: these include changes to the contractual position which restricted the ability of the Tier 1 manufacturer to appropriate value from product innovation.

- **Changes to the nature of knowledge**: these changes affect the nature of product knowledge, which was codified into the product lifecycle management system (PLM) owned and controlled by the OEM.

Digitalization of product development and manufacturing execution

In this case study, the data relates to a collaboration project aimed at new product and process development, followed by a contract for the manufacture and supply of a new product utilizing new manufacturing technologies. This project took place over 32 months and involved eight parties from across the automotive manufacturing value chain:

- The OEM
- A Tier 1 manufacturing (lead party)
- A technology provider
- Two Tier 2 manufacturers
- Three research institutes

The objective of this project included the development of a new product which forms part of the new generation of electric vehicles. As a consequence of the novelty of the product and the vehicle in which it is going to be utilized, this project also aimed to develop a novel manufacturing technology in order to enable the manufacturers involved to manage the high levels of product complexity surrounding manual process and high-precision manufacturing techniques.

In the same way as in previous case studies, this project was governed by a collaboration agreement that established the roles and responsibilities of each of the parties in the project, as well as the rules surrounding background and foreground IP.

This project has utilized the latest digital technologies for the product and process development. These technologies allowed the collaborating parties to exchange product- and manufacturing-related data in order to virtualize the product and manufacturing processes. As a result, the value chain as a whole was able to accelerate the product and manufacturing development cycle by running concurrent product design, design for manufacture, and multiple product and process simulations.

A number of digital models, also referred to as digital twins, were created to characterize and simulate the product and manufacturing performance with a high level of fidelity to the physical process. These models aggregated knowledge from multiple departments and disciplines across the collaborating parties. For example, the operations team and the manufacturing engineering team provided real-life operational data to inform the simulation models utilized to refine the product design.

In a similar way to Case Study 2, on the successful completion of this project, the Tier 1 manufacturer was awarded the contract for the supply of the product to the OEM for a period of 18 months. Subsequently, two of the Tier 2 suppliers were also awarded the contract for the supply of components to the Tier 1 supplier.

According to the data analysis of this case study:

- The Tier 1 manufacturer is now in the weak quadrant, with a lower level of IP protection at 3.7, compared to 4.7 in Case Study 1, and with a lower level of inherent replicability scoring 3.5, compared to 5.5.

- The OEM position has improved and remained in the strong quadrant scoring 8.4, compared to 6.6, for IP, and the inherent replicability score has also increased to 6.3, compared to 5.5.

- In contrast to Case Study 2, the position for the technology partner has also improved in regard to the IP score and decreased in regard to the inherent replicability scores. Whilst remaining in the strong quadrant, the IP increased to 9.3, compared to 8.4, and inherent replicability has decreased to 5.6 in comparison to 7.1 in Case Study 1.

- The university partner's position also remained in the strong quadrant, maintaining the same score as Case Study 2 with IPR at 9.7, which is a decrease compared to 10 in Case Study 1, and also a decrease in inherent replicability to 5.0, compared to 6.1 in Case Study 1.

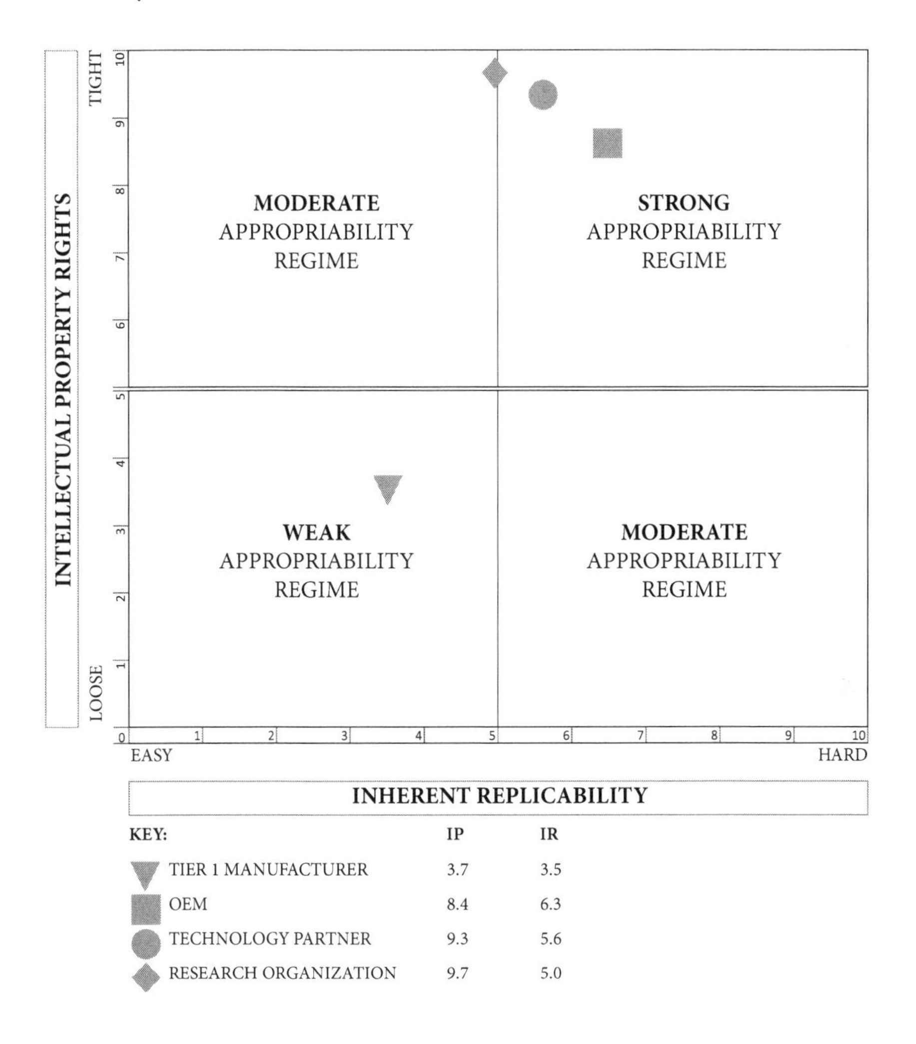

Figure 3: the position of each partner in Case Study 3 (digitalization of product development and manufacturing execution) according to the Marc Model analysis

In Case Study 3, the appropriability position of the Tier 1 manufacturer is further weakened by the intensive use of Industry 4.0 technologies across product and process development and manufacturing stages. Whilst on the other hand, the position of the OEM continues to increase in strength. The main influences on this shift are:

- **Changes to the contractual position**: new clauses were included in both main agreements controlling the relationship in the collaboration project and the supply chain contracts.
- **Changes to the nature of knowledge**: the production knowledge was codified into a single system and a large amount of data from both product and process simulations was shared with all the partners.
- **Changes to the practical means of protection**: these changes included uncontrolled operational data exchanged and aggregated in a digital collaboration platform used to simulate product and manufacturing processes in parallel.
- **Changes in the level of generic competencies**: these changes are related to the technical competencies required to manufacture the new product. In this particular case, the new manufacturing method required generic competencies that can be found in the majority of suppliers in the automotive value chain.

Digitalization of manufacturing execution for current products

This case study explored and collected data regarding a collaboration project that was aimed at a new process development, followed by a contract for the manufacture and supply of an existing (already developed) product utilizing new manufacturing technologies. This project took place over 24 months and involved five parties from across the automotive manufacturing value chain:

- The OEM
- A Tier 1 manufacturer
- A technology provider (collaboration lead party)
- A Tier 2 manufacturer
- A research institution

The aim of this project was to develop a new manufacturing technology capable of achieving a higher level of productivity compared to previous methods to make a product already under contract for its manufacture and supply between the OEM and the Tier 1 manufacturer.

The project was governed by a collaboration agreement which established the roles and responsibilities of each of the parties to the project, as well as the rules surrounding background and foreground IP. However, in this case there was a contractual relationship already in place between the OEM and the Tier 1 manufacturer which governed the ownership of IP.

The collaboration project extensively used the latest digital technologies for manufacturing simulation at machine, production-cell, production-line and plant levels. These simulation models were used across the five collaborating parties to exchange data, accelerate the development of the manufacturing process development and de-risk capital investment in manufacturing equipment through virtual commissioning and digital validation.

Upon the successful completion of this project, the Tier 1 manufacturer was awarded an extension to the contract for the supply of the product to the OEM for a period of three years. The Tier 1 manufacturer was also awarded a contract to supply the same product to another continent for the extension period of three years. Subsequently, the Tier 2 supplier was awarded the contract for the supply of components to the Tier 1 supplier.

According to the data analysis of Case Study 4:

- The Tier 1 manufacturer is now at the weakest position in comparison to the other case studies, with the lowest scores on both IP protection at 3.5, compared to 4.7, and the level of inherent replicability scoring 2.7, compared to 5.5 in Case Study 1.
- The OEM position is at the strongest point across the case studies with an IPR score of 9.0, compared to 6.6 in Case Study 1, and an inherent replicability score of 6.6, compared to 5.5 in Case Study 1.
- Contrary to the OEM, the technology partner position has suffered a detriment on both scores. Whilst still positioned at the strong quadrant, the IPR scores decreased to 8.0, compared to 8.4 in Case Study 1, and inherent replicability has decreased to 5.2, in comparison to 7.1 in Case Study 1.
- Lastly, the university also suffered a detriment to its position and a decrease in both scores with IPR at 9.0, compared to 10 in Case Study 1, and inherent replicability at 4.8, compared to 6.1 in Case Study 1.

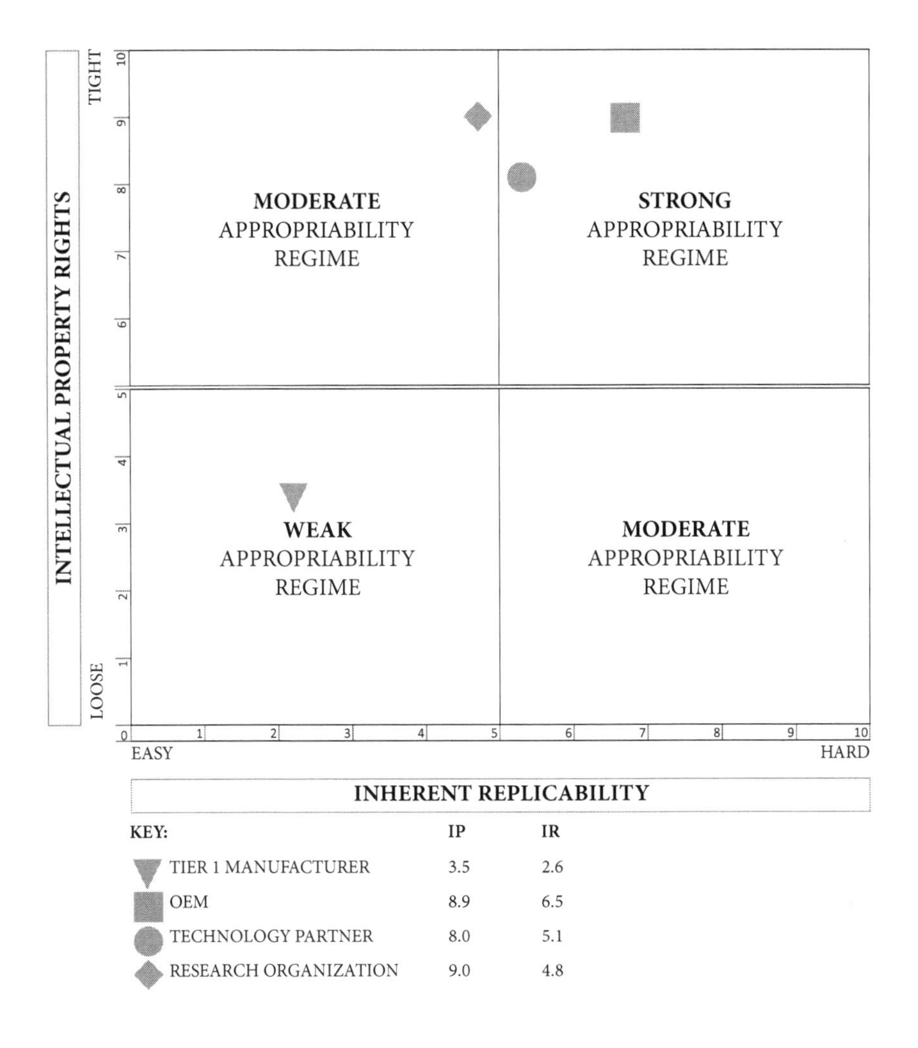

Figure 4: the position of each partner in Case Study 4 (digitalization of manufacturing execution for current products) according to the Marc Model analysis

In Case Study 4, the position of the Tier 1 manufacturer is further weakened, whilst the position of the OEM reaches its strongest point. Similar to Case Study 3, the main influences on this shift were:

- **Changes to the contractual position**: new clauses were included in both main agreements controlling the relationship in the collaboration project and the supply chain contracts. One of these changes imposed significant barriers to the value appropriation on the Tier 1 manufacturer. The most significant of these changes imposed an obligation on the supplier (the Tier 1 manufacturer) to assign all IP in relation to both product and process to the customer (the OEM).

- **Changes to the nature of knowledge**: the production knowledge, as well as the new operation data, was codified into a single system and a large amount of data from the Tier 1 manufacturer operations was made available to the OEM and technology partner.
- **Changes to the practical means of protection**: these changes included uncontrolled operational data exchanged and aggregated in a digital collaboration platform. It also included the live operational data, including investigations and reason codes for process-related issues.
- **Changes in the level of generic competencies**: the product produced in this particular case is well known to the automotive manufacturers and is commercialized as a commodity. The competencies required to imitate it are widely available in the automotive manufacturing value chain.

10.
COMPETITIVE ADVANTAGE IN DIGITAL MANUFACTURING

By taking the four case studies of Industry 4.0 collaborations together, we can start to see the overall trends where gains and losses are being made across value chains. In this chapter, all the businesses participating in the case studies are plotted on a single figure (see Figure 1) to compare the impact of Industry 4.0 on their competitive positions. This combined view of the Marc Model provides a unique perspective of transformation in the manufacturing appropriability landscape which results from the introduction of digital technology.

At a glance, the model demonstrates that apart from the original equipment manufacturer which invariably improves its appropriability position in all case studies, all other parties are in a weaker position due to the adoption of digital technologies and the integration of the value chain in these projects. Let's now review each quadrant in the Marc Model and the implications for the businesses in those positions.

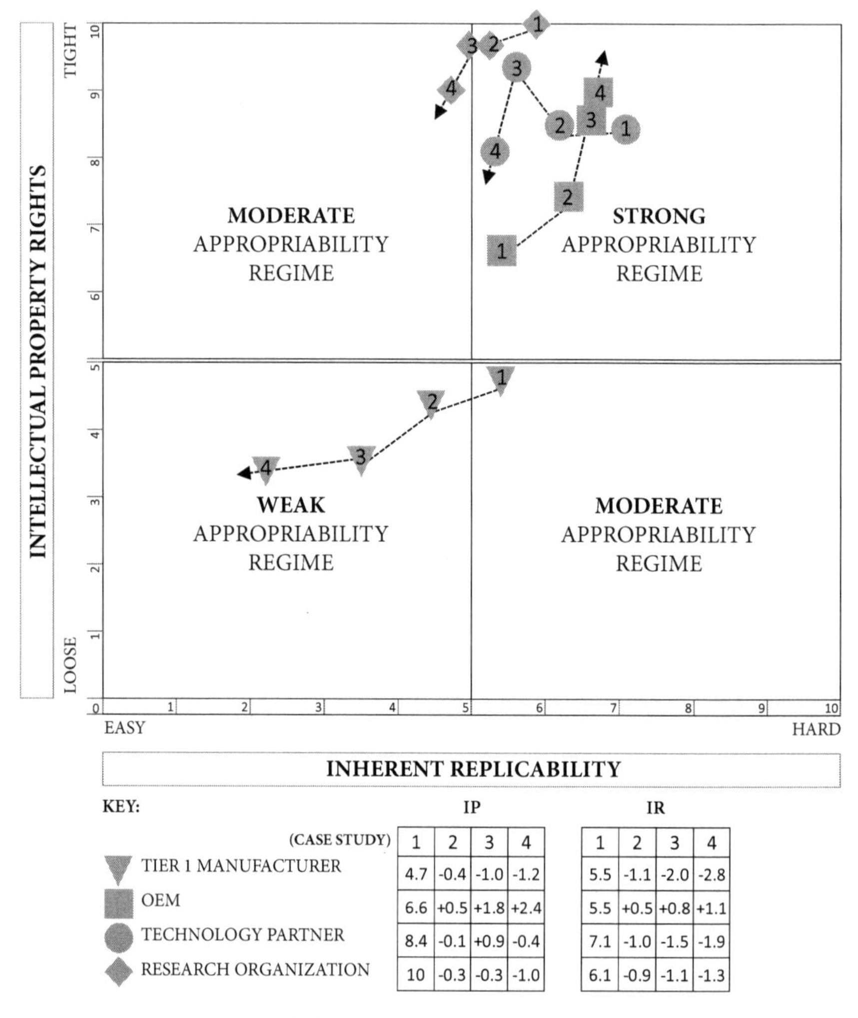

(CASE STUDY)	IP 1	IP 2	IP 3	IP 4	IR 1	IR 2	IR 3	IR 4
TIER 1 MANUFACTURER	4.7	-0.4	-1.0	-1.2	5.5	-1.1	-2.0	-2.8
OEM	6.6	+0.5	+1.8	+2.4	5.5	+0.5	+0.8	+1.1
TECHNOLOGY PARTNER	8.4	-0.1	+0.9	-0.4	7.1	-1.0	-1.5	-1.9
RESEARCH ORGANIZATION	10	-0.3	-0.3	-1.0	6.1	-0.9	-1.1	-1.3

Figure 1: comparison of the position of each manufacturing partner in four appropriability case studies according to the Marc Model analysis

The weak appropriability quadrant

As demonstrated in the model above, Tier 1 manufacturers are typically in the weak quadrant. Businesses in this position are normally mass producers of products without trade marks or standard products in an industry where imitation is easy. These businesses typically have no property rights on either the product or the processes for manufacturing the product.

Manufacturing businesses in this field typically become producers of commoditized or standard products or services which are not protected. This reduces the chances of appropriating value and investment in innovative products or processes.

The first moderate appropriability quadrant

In the bottom right of the Marc Model, apart from the Tier 1 manufacturer in the baseline, there are no businesses positioned in the first moderate field. An example of a business that is positioned in this quadrant would be IKEA, the Swedish furniture giant which has a unique distribution and logistics system. Businesses in this quadrant typically also have an extensive knowledge of their customers and draw on this knowledge continuously; they are in a constant learning process and as a result, even though their IP rights protection is weak, other organizations find it difficult to imitate their processes.

The second moderate appropriability quadrant

As the model demonstrates, the university partners are positioned in the second moderate quadrant (top left) in Case Study 3 and Case Study 4. This position is typical of businesses that have strong IP processes and protection through rights such as patents and trade marks that protect them from imitation by competitors. Nevertheless, their products or technologies are relatively easy to replicate, which appears to be a change in the case of university partners in scenarios such as Case Study 3 and Case Study 4 where vast amounts of information are aggregated and shared across multiple entities.

The strong appropriability quadrant

In the top right of the Marc Model, the technology partners and the OEMs are typically positioned in the strong quadrant due to strong IP processes, as well as strong research and development. It is also difficult for other companies to copy their ability to develop new products, processes and competency areas. Additionally, OEMs and technology partners maintain this position by locking collaborators and partners into strong contracts where the ownership of IP in relation to products and processes is typically controlled by them.

Maintaining and developing competitive advantage for Industry 4.0

The appropriability regime described above combines the different mechanisms to maximize value capture for return on investment for a particular innovation. The choice of protection mechanisms should be aligned to the business strategy and be unique for each manufacturer depending on their particular business model, value-chain relationships and industry. It was evident through the data analysis that the interviewees have a limited understanding of how different protection mechanisms could support their strategies. The same was true in terms of efficacy (strength) of the mechanisms. As described by a Tier 1 manufacturer director:

> We have been involved in over ten collaboration projects over the past four to five years, but there is nobody looking after IP. I think that the university partner has registered a couple of patents, but we didn't get anything.

This limited understanding was the case even with one of the OEMs interviewed as demonstrated by a senior engineer who stated that:

> We only use patents to protect our inventions, so that if a competitor tries to imitate our products, we will go after them and make sure we stop it. The thing I am not sure of is whether we can identify people using or imitating our product.

Another interesting remark regarding IP knowledge demonstrated that the lack of understanding extends beyond the operations and engineering teams to support services who in the past were seen as the IP gatekeepers in some businesses as stated by the managing director of a Tier 1 manufacturing business:

> We definitely need to increase our knowledge of IP law. I feel like the lawyers lack our manufacturing knowledge in operations and commercials, so it is difficult to apply the IP protection law in isolation.

These case studies demonstrate that as Industry 4.0 technologies are implemented across the manufacturing value chain, there is an increase in businesses collaborating and exchanging data regarding different technologies. In these scenarios, predicting the innovation project outcomes and the best IP strategy becomes more complex. Additionally, the protection mechanisms used by manufacturers are also likely to increase in complexity. In our interviews, this point was made clear by the commercial director of a Tier 1 manufacturer:

> I think most of our team would agree that with digitalization it will become more difficult to protect our margins and the value in the current products, as the customers will be able to see everything; there will be nowhere to hide any bounce.

A similar point was also made by the operations director of a Tier 1/2 manufacturer:

> One way to look at it is that further connectivity with our suppliers and customers will enhance our relationships. The other is that the customers could bypass us completely and go straight to the suppliers themselves. This is not new in automotive manufacturing, but digitalization could increase it.

To address this complexity, further combinations of different protection mechanisms should be developed by manufacturers. These could include for example a combination of data secrecy, which is an attempt to manage key information availability on a need-to-know basis, and improved contractual terms and conditions, which prevent knowledge sharing by unauthorized personnel.

The protection mechanisms utilized by manufacturers should include a mix of prerequisite, derivative and supportive appropriability mechanisms. The mechanisms known as prerequisites, as the name says, are those which are required to enable other applicable protective mechanisms. An example of a prerequisite mechanism is that in order to patent an invention, the inventor must ensure that the invention retains its novelty by protecting any early disclosure of the invention to external parties who are not bound by a non-disclosure agreement. Derivative mechanisms buy some time in order to derive an appropriate strategy and also to make use of competitive advantages such as shorter lead time to market. Using patents as an example again, this protective mechanism provides a monopoly protection for a manufacturer whilst the R&D teams can work on the next wave of innovation for products and services required to derive a competitive advantage for the business.

The protection mechanisms known as 'supportive' combine with other formal and informal mechanisms, such as the use of contractual agreements and IP training, to support employees in their duties and responsibilities regarding confidential information which may lead to issues with patent applications and trade secret enforcement.

The selection of protection mechanisms and IP strategy will provide a level of protection by raising the barriers for competitors to imitate. There are also other benefits associated with strategic opportunities for value appropriation, such as the potential to generate revenue from licences of protected technologies and other methods of monetizing the data generated in the horizontally integrated value chains.

There is evidence from the interviews that parts of the value chain are aware of the need to develop IP capabilities and adapt their IP strategies as evidence by the following comments.

> Our team is going to be trained in IP as a matter of urgency. We cannot afford to have engineers sharing our latest inventions with the customers; we all lack education in this area (engineering director, motorsport technology provider).
>
> There are businesses out there using IP as a currency in kind to enter into collaborations and even joint ventures, but most of the Tier 1 manufacturers we work with in the UK are missing this opportunity (senior IP manager).
>
> I cannot stress enough how important IP is for our business today, as it is the main reason we form our partnerships and joint ventures, but even more so in the future where everything we do will be digitalized and available to our customers (engineering manager, Tier 1 automotive manufacturer).

Manufacturing businesses should also consider the efficiencies and savings of selecting the right protection strategy, as the selection of inappropriate mechanisms may lead to high costs. Patents, for example, take time and resources throughout the application process and, even after they are granted, can result in expensive legal bills related to enforcement cases. The same logic applies to financial incentives for key personnel, which may preclude employees from leaving with knowledge in the short term, but will also affect the manufacturer's profitability.

This perception of IP as costly was evident in the interviews as pointed out by a senior engineering manager at a Tier 1 manufacturer:

> The costs of IP protection are too prohibitive, and it also takes too long. I am not sure whether we would even be able to protect anything with our current contractual agreements and no budget.

A similar point was also made by the managing director of a Tier 1/2 manufacturer:

> Yes, we own IP. I think we own three or four patents and trade-mark protection on our name, logo and branding. There is not an awful lot of IP and it is expensive to protect and maintain. Due to costs, our strategy is to only protect IP as a last resort and we survived over the last three decades just fine.

An important distinction for manufacturers to make when selecting the right protection mechanisms and setting their IP strategies relates to the distinction between incremental and radical innovation, which is critical to ensuring efficacy.

Typically, protection mechanisms and IP strategy for closed and incremental innovation that improves upon existing technologies in processes, products and services must be stronger than those utilized for the protection of radical innovation. This difference is because existing technologies tend to be easier to imitate by competitors and require a short time to commercialize due to the competencies in the value chain, and the level of confidence and acceptance by existing customers. The opposite is true for radical innovations which are not yet proven and more complex to replicate. In some cases, it is even an effective strategy to encourage sharing under certain conditions, as the wider adoption of the radical innovation can result in profits to the manufacturer.

Certain technology acquisition strategies appear to be more frequent and relevant than others. More specifically, the interviews reveal a large dependence on different means and types of collaboration for adding digital technology to their technology base. This high dependence on collaborations can be understood through the highly distributed areas of technology that have to be combined before they can be applied to manufacturing products or operations. In addition to supporting the acquisition of skills, collaborations are highly relevant in multi-invention contexts such as those brought about by the digitalization of manufacturing.

The findings from the case studies demonstrate that there is a high level of uncertainty amongst the manufacturers regarding these collaborations, particularly in relation to issues of IP allocation. These included, for example:

- Who gets the resulting IP on what?
- How is the background IP going to be compromised in the exploitation of the resulting IP owned by another party?
- How can secrecy can be ensured from partners?
- What can partners do with the knowledge they acquire?

The interviewees have shown evidence that the typical processes and IP management in manufacturing were designed to address simple relationships between two commercial partners. In such relationships, it is reasonable to assume that both parties want ownership of at least some IP created to ensure the appropriability of value from the innovations that they intend to commercialize. This finding is aligned to the literature in regard to pre-digitalization and open innovation industries.

These findings point out that in order to ensure appropriability, openness in terms of developed knowledge needs to be limited. In turn, in order to limit the risks, a range of protective mechanisms which goes beyond secrecy is needed, ie, more complex contracts are required to govern the IP allocation between the parties.

On the other hand, collaborations require a certain openness between the involved parties in order to achieve the project objectives and for there to be a fruitful exchange of knowledge and capabilities. It could thus be said that manufacturers will have to be more aware of the strategies of each potential collaborator, the risks and benefits of exchanging knowledge and capability, and the need to balance their openness and appropriability from innovative outcomes in collaborations.

Nevertheless, some of the manufacturers interviewed are not

intending to change their approach to IP as evident in the comments from Tier 1 manufacturers.

> I don't think we will be adapting our strategies; the law is not changing, so in a way it will be the same rules. As always we will strive to comply with the OEM's requirements and its terms and conditions (managing director, Tier 1 automotive manufacturer).

> We don't have an IP strategy as such; it is typically the business strategy that guides everything else, so if a particular business generates some IP, we will try to protect it (financial director, Tier 1 automotive manufacturer).

These issues are increasingly relevant as manufacturers engage in higher levels of collaborations to develop innovations in products and processes that involve digital technology. The roles of IP rights and a comprehensive IP strategy are an important enabling factor for collaborations between multiple partners and to achieve the levels of value-chain integration required by Industry 4.0. Nevertheless, most of the manufacturers interviewed have a limited IP function within the organization and only rarely was this function observed to have a strategic role.

In almost every case, the IP function operated as a support function instead of an integrated part of the overall business. The protection mechanisms often focused on differentiating aspects of the business models, and the IP department had the responsibility of executing the direction of the business leaders and functional leaders.

The empirical data from this study also demonstrate that the IP landscape and thus the appropriability regimes for manufacturers will be impacted by Industry 4.0 adoption. This was evident in the shift towards a weak appropriability regime for Tier 1 manufacturers as demonstrated as part of the Marc Model analysis.

Re-aligning strategy and practice

Findings from the interviews demonstrated that IP is an increasingly important part of manufacturing business in the context of Industry 4.0. However, most interviewees have also shown a huge disconnect between their answers and their business practices identified through their contractual agreements and businesses structures. Also, even if the IP functions in manufacturing businesses view IP increasingly as an integral part of business strategy, there is no evidence in the contracts of those interviewed that these strategies are being aligned or integrated in any form.

This perspective also helps to elucidate the extent to which the appropriation regime for manufacturers will be impacted and how the understanding of IP strategies can be adjusted to protect innovation in the context of Industry 4.0.

When viewed from a business model perspective, manufacturers are faced with types of relationships, collaborations, contracts and networks where an unprecedented level of data will be exchanged across different businesses in the value chain. As a result, these relationships and collaborations lead to a new work environment and transform the work practices in the automotive manufacturing value chain.

These changes are important in regard to IP strategies, not only because current manufacturing business are set up according to a pre-established framework, where the relationships are governed by contractual agreements, standard practices and procedures, which control interactions in the value chains, but also due to the fact that unprecedented levels of integration and data exchange will bring transparency to the value chain to a level never seen before.

The following figure shows a comparison between a simple representation of the current manufacturing value chains and the future Industry 4.0 connected value chains. On the left hand side, the figure shows a typical value chain where there is limited interaction and data exchange between businesses in multiple tiers resulting in 16 connections representing the current relationships of the 16 businesses in the value chain. On the right hand side, the figure shows the potential effects of a horizontally integrated value chain where there are multiple relationships with data and information exchange across all businesses in the value chain resulting in 256 relationships in the same value chain.

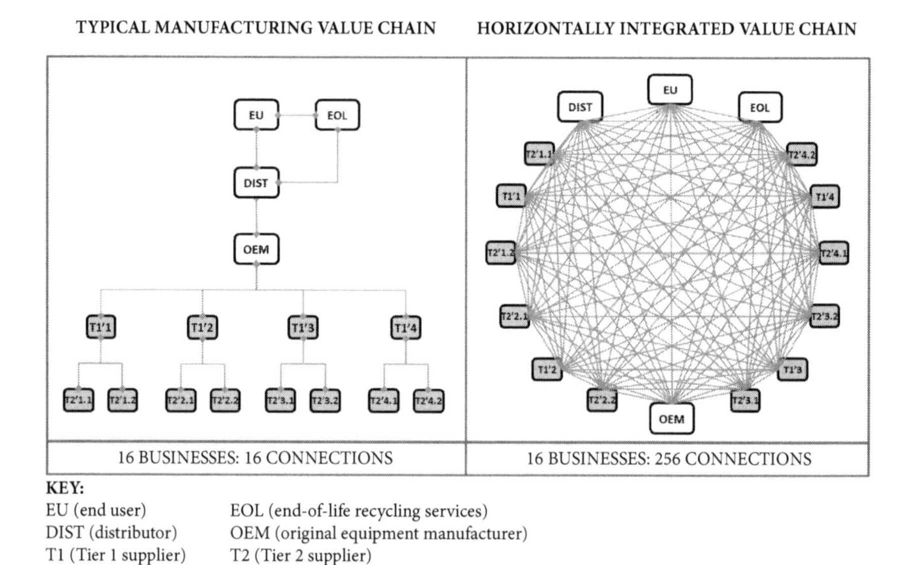

Figure 2: manufacturing value chain comparison

The effects of digitalization and integration in the automotive manufacturing value chain affecting the manufacturers' relationships should be carefully considered. An important area for focus is the standard contractual agreements, such as contracts for the supply of services and products, collaboration agreements and confidentiality agreements. These should be reviewed and their relevant terms and conditions adjusted to account for the new ways in which each business collaborates, and how it exchanges and uses data across the value chain.

These contractual agreements should provide a level of protection to current and future businesses relationships, where, for example, products are developed and made collaboratively with multiple businesses across the value chain. These contractual agreements should also consider where services are sourced via shared platforms and where most of innovation and operational data are stored in a shared cloud.

This transformation results in a number of challenges to IP strategy and management, as IP practitioners working in manufacturing have historically used IP rights in the traditional sense to protect the physical things, devices, structures and even the configuration of inventions embodied in physical systems, physical outputs, the operation of physical systems etc.

However, with the implementation of Industry 4.0, the focus needs to be expanded to the IP protection of intangible things such as methodologies, virtual systems and its configurations, data ownership, data handling, data storage, algorithms, data sets, databases, brand, brand recognition etc.

The digital transformation resulting from the implementation of Industry 4.0 challenges the current understanding and use of IP protection and commercialization strategies in manufacturing. This change requires the development of new approaches that will be better suited to the rapidly changing, highly integrated business networks.

Such a position was clearly made in the *Made Smarter Review* in 2017, which recognizes the importance of IP as a key intangible asset that can make up over 80 percent of the value of a manufacturing business (Visual Capitalist, 2020[1]) and the fact that IP is often the key to securing a competitive advantage in globalized manufacturing value chains.

The review was commissioned by the UK government and led by Professor Juergen Maier, the former chief executive of Siemens UK, who also recognized that IP theft is one of the key threats related to the digitalization of businesses (*Made Smarter Review*, 2017[2]) . The review also points out that due to the intangible nature of IP, which is typically found in digital information, it is susceptible to digital piracy.

As the findings from the data analysis demonstrated, with the implementation of interconnected communications and the increase in manufacturer collaboration, manufacturing businesses are faced with the challenging task of carefully considering how to protect their IP, whilst at the same time how to facilitate interoperability between businesses in the value chain.

Having discussed the impact of Industry 4.0, horizontal integration of manufacturing businesses and appropriability regimes, attention will now turn to the recommendations to improve the potential value capture and profits for manufacturers.

Notes

1. Report available at https://www.visualcapitalist.com/intangible-assets-driver-company-value/
2. *Made Smarter Review* (2017), available at: https://assets.publishing.service.gov.uk/government/uploads/system/uploads/attachment_data/file/655570/20171027_MadeSmarter_FINAL_DIGITAL.pdf

PART 3

PROFITING FROM INDUSTRY 4.0

11.
YOUR INDUSTRY 4.0 PROJECT

Following the first two parts of this book the question that remains unanswered is how can Industry 4.0 help your business be more profitable? This is in fact the key question to be addressed in Part 3, which begins with a practical guide to getting started with Industry 4.0. Do not worry if you are already further down the digitalization journey as there is plenty of helpful information in the other five chapters of Part 3 to help you improve your chances of profiting from Industry 4.0.

As you have hopefully gathered by now, Industry 4.0 in isolation will not deliver substantial value for your business. Industry 4.0 technologies and solutions are rather a way to generate value through new or existing products, processes and product-related services that enable you to reduce costs, increase sales and improve profitability by offering more value to your customers.

There are plenty of solutions and suppliers of Industry 4.0 products and services out there. However, to achieve the potential benefits, you will need a unique combination of technologies and solutions designed to suit the needs of your business. This chapter will provide a guide to help you and your business successfully select technologies, develop value-driven Industry 4.0 innovative applications and effectively launch them into action.

This gradual process of Industry 4.0 adoption is likely to involve fundamental changes in the way your business operates. These include changes in operational processes, as well as transformation of business models. Due to this organization-wide change, your business decisions and strategy for Industry 4.0 adoption need buy-in across the entire organization, particularly among senior managers who can make the commitments to allocate the required resources to ensure the delivery of your digitalization projects.

Setting your project structure

The first action to take in this journey is to form a suitable project team. Ideally, it will be formed by an interdisciplinary cross-departmental group consisting of staff from every area and level of the business. This is especially vital when generating ideas for innovation which requires close collaboration between information technology, operational technology and engineering. This team will then help your business navigate through the suggested activities based on the four steps that this chapter outlines. These activities can be planned and scheduled over a suitable period of time which works for you. They could, for example, take place over a couple of weeks of intense workshops or they could be organized over a more extended period with activities in shorter bursts over six months.

Regardless of what approach you take and what resources you have available, the generic practical steps to follow have been tried and tested with many manufacturers, so should help you get started and get your team on the same page as to how Industry 4.0 can add value to your business. These steps are not exhaustive, but rather a set of guidelines that can be executed by your Industry 4.0 inhouse project team with minimum external support required.

The four steps begin with an analysis phase capturing and exploring your business's areas of expertise and key competencies. Building upon this initial analysis and the opportunities highlighted, the second step can be initiated. This can be referred to as the ideation phase where the team will generate ideas relating to products, processes and services that can be enhanced by Industry 4.0 technologies. This can take place via a workshop supported by Industry 4.0 technology experts (internal or external as required) and the project team to help elaborate and assess the potential ideas. This is an opportunity to level the playing field for all involved. The dissemination of the analysis, particularly the gaps and opportunities, can and should be shared with the whole team, before moving onto the development of new ideas for operational and business models.

In the next step, which can be referred to as the business case phase, the project team will then conduct a thorough evaluation of these solutions and concepts with regard to their potential value-add impact and the resources required for taking the ideas into Step 4, the implementation phase. In this final phase, a full implementation plan is created and executed, and your analysis data is updated to reflect your new competencies.

Analysis phase

The objective of Step 1 is to identify the core expertise and competencies currently available in your business in the context of Industry 4.0. In this effort, you should also consider your value chain, the market positioning of your business and its level of expertise regarding Industry 4.0 technologies in comparison to your partners and competitors, whether actual or potential. This analysis should include the creation of a benchmark ranking your business's competencies or expertise in technologies, processes, products and services.

It's critical to identify the current status of your Industry 4.0 capabilities, so you can create and evaluate ideas in the next two steps. The Step 1 analysis will result in a business-wide view of existing competencies that will form the basis for your business's future skills and capability development plan.

Achieving a business-wide, cohesive and honest understanding of the existing competencies is the cornerstone of a successful and profitable introduction of Industry 4.0 technologies into your business. During the ideation phase, your Industry 4.0 project team will develop ideas based on this common understanding.

This competency analysis can be divided into internal and external considerations and should focus on two questions:

- What are the current areas of Industry 4.0 competencies in the business?
- How do the existing competencies compare to internal and external perceptions of your business?

The internal analysis of competencies can reveal your hidden Industry 4.0 competencies and provide a holistic view of your current capabilities. Manufacturing businesses usually possess Industry 4.0 competencies and specialist knowledge in certain areas. However, these are not normally categorized and shared within all areas and levels of the business. In this analysis, you should identify and include all existing activities, approaches, technologies and strategies across the business. This list will be critical to evaluating the potential ideas generated in Step 2, as well as estimating the internal and external levels of effort and resources required to implement the potential ideas.

In the external analysis, you will explore how your Industry 4.0 competencies are perceived by your value-chain partners and customers. It will give you a baseline for what image you have for Industry 4.0. You can then judge how attractive your brand and reputation might be for potential partners. It can also guide your future negotiations about how to distribute intangible assets across new value chains you are involved in forming.

External perspective

The objective here is to explore and document your business's external offering and the portrayal of its Industry 4.0 capabilities across your value-chain partners and customers. This analysis will provide an indication of the extent to which your business has already built a reputation for Industry 4.0 competency.

The analysis is not based on exploring internal projects and capabilities, as these are normally not visible and open to the value chain and to customers. In this case, you are rather interested in the outsider's perspective, which include your presence on social media, in press releases, thought leadership pieces and membership of bodies.

The external competency profile can be used by the project team in workshops designed to increase awareness of the current external profile, discussing areas of weakness or those of future importance. As the project team will be formed from departments across the business, the members are likely to possess detailed knowledge of the business process, product and service offering, so will be able to perform an analysis of areas of development to improve external perception of Industry 4.0 capabilities in your business.

The assessment of both internal and external competencies can be split into categories which you can assess against a Likert scale of five levels, which is typically configured so that level 1 represents no competency and level 5 represents leading competencies in your business. It is up to you how you categorize levels on the scale. The point is to codify the competencies in a commonly understood method so that you can gain a common understanding across the business.

The categories to be assessed can also be tailored to suit your business. In the past we used categories based on Industry 4.0's nine key technologies. For example, the assessment categories to get you started could be a technology competency analysis:

- Big data and analytics
- Autonomous robots
- Simulation
- Horizontal and vertical systems integration
- Industrial internet of things
- Cybersecurity
- The cloud
- Additive manufacturing
- Augmented reality

Internal perspective

In my experience working with small, medium and large manufacturing companies across the value chain, I have found that the current product and service offerings often do not reflect the entire range of competencies available within a business. The internal competency analysis helps identify your Industry 4.0 capabilities beyond those in mature products and solutions being commercialized with a high technology or manufacturing readiness level. You should pay particular attention to the identification of these untapped intangible assets and ensure that the whole organization, particularly the members of your Industry 4.0 project team, understand exactly what these unused competencies are in order to inform Steps 2 and 3 in this process.

The approach to identify these competencies can be based on an analysis of each department. It is common to find that different teams will have distinct levels of Industry 4.0 maturity. For example, product design and development tend to have more experience in complex simulations and digital twins, whilst manufacturing and production engineering may have mature capabilities on robotics and data analytics, and IT teams tend to be more proficient in areas such as networks, big data, analytics and cloud computing.

Ideally, your internal capability analysis should be built using data from participants representing an interdisciplinary cross-section of your business to identify the overall competency profile. You should include at least one senior manager and one team member from each area in the data collection to capture a representative sample size.

The data analysis can be consolidated and presented as a radar chart demonstrating the capabilities levels and Industry 4.0 maturity in relation to products, processes and services or in relation to the categories you have selected for the assessment. It is recommended that the analysis results should be shared with the whole business in order to create a common understanding of existing capabilities, as well as clear areas where there are gaps and opportunities for development. This cross-organization, interdisciplinary discussion will create a strong understanding of your Industry 4.0 competencies.

Ideation phase

The objective of this step is to generate ideas and create new concepts for solutions and new business models capable of addressing current and future disruptive changes and/or opportunities for your business. The execution of this step is dependent on the analysis phase and typically takes place in two stages.

In the first stage, the project team can run a workshop to identify and collect initial ideas to utilize Industry 4.0 solutions. In the second stage, these can be discussed and further developed into propositions. At the end of this period, these Industry 4.0 propositions can be developed into full 'solution concepts' to enhance value to current business models or even to create entirely new business models.

In the creation stage, the team will focus on generating as many ideas as possible in regard to new products, services or operational processes by utilizing Industry 4.0. In the development stage, the team can be split into groups based on alignment between ideas and capabilities to further develop these ideas into solutions.

Creating ideas

This stage can begin with the generation of ideas by individual participants. To focus the effort, the ideas can be categorized into broad areas, such as ideas for products, services or processes. These ideas should prioritize the key areas where gaps and opportunities were highlighted as part of the competency analysis in the previous step. In particular, the participants should be encouraged to generate ideas with the potential to aggregate new value and delight your customers.

The criteria for aggregating value or delighting customers are a moving target. Solutions and product features that aggregate value today typically evolve and become basic requirements in the future. It important to categorize the ideas into a pipeline of continuous innovation to maintain competitive advantage.

This process typically takes place as part of a workshop and begins with individual participants being asked to generate ideas and enhancements for existing products, services and processes based on Industry 4.0 technologies and design principles. Maturity levels can be assigned to each of the products, services and processes, and the ideas generated can be assessed in order to identify how the maturity levels could be improved.

At the conclusion of this first step in generating ideas, each of the participants present their ideas to the rest of the group. A simple one-page template can be used to help the participants capture and present the key content of the idea, the current maturity level of the relevant product, service or process, and the potential impact of the idea.

Developing ideas

Following the individual generation and presentation of ideas, the workshop facilitators support the team to cluster the ideas into areas or topics which will form the basis for splitting the workshop participants into specialized groups based on the alignment between the ideas and their expertise. These specialist groups will work together to develop, expand and refine the ideas (mental constructs of possibility/opportunity that someone suggested) into solution concepts (forms of ideas that have gone through the process of fine-tuning and considering how the ideal can be applied as a solution to address a specific problem or opportunity).

The development of the ideas can take place in two distinct steps. First, the ideas can be prioritized depending on their potential to aggregate value to the business. It is important that the concept of value used in this prioritization is broader that just looking at short-term revenue. The criteria of value has to be contextualized to the particular business model of your organization.

For example, an operational improvement idea to deploy robotics to improve a particular process may deliver short-term improvement to your business. However, the development of a new product or service offering may deliver long-term prospects of an entirely new revenue stream to support your business strategy. Whilst prioritizing the ideas, the team can also consolidate similar and complementary ideas into stronger solution concepts.

In the second step of idea development the groups are asked to perform a business model exploration of the top two or three solution concepts according to their prioritization. I would like to suggest that the groups should use the St. Gallen Business Model Navigator or equivalent tool to support this exercise which typically focuses on answering four questions[1]:

- **Who**? Who are the target customers (internal and external) for the solution concept?
- **What**? What is the offering to the targeted customers?
- **How**? How is the value proposition created?
- **Value**? How is value/revenue generated?

Exploring each of the solution concepts by answering these business model questions will help the groups to take a systematic approach and consider the potential business value, instead of just considering how good the ideas or technologies are. This approach also helps to move from solution concepts to business-model concepts which will help in the creation of a business case for investment.

To conclude the idea development, the project team can record the results of the analysis performed by each group in a standardized manner to facilitate the sharing of these solution and business-model concepts with the wider organization in advance of the kick-off of Step 3, the business case phase.

Example of ideation in practice

A typical first step in manufacturing digitalization involves activities supporting the instrumentation of analogue operational processes and machines that do not possess any digital data capture and exchange capability. In this scenario, the idea generation activity begins by selecting a particular process and creating a map characterizing each step of the process, including any associated IT and OT equipment. Each of the steps are assessed and a digital maturity level is allocated.

Once this is completed, the team begin to generate ideas to improve the digitalization of each of the steps. This could involve the introduction of data collection devices or passive data memory for the purpose of capturing production data for future analysis using traditional statistical methods or even artificial intelligence, such as machine learning and other forms of advanced analytics.

This idea could be described as the implementation of an IoT device for capturing, processing, storing, analysing and sharing data from a process or machine in order to provide operational insight and support manufacturing improvement activities. Developing it into a solution and business-model concept would begin with the following considerations:

- **Who?** Who are the target customers (internal and external) for the solution concept? Potential internal customers: operational team, quality Team and senior management. Potential external customers: customers of products and services provided as a result of the processes being digitalized and potential manufacturers seeking to digitalize similar processes.
- **What?** What is the offering to the targeted customers? For potential internal and external customers: physical IoT device and services related to the data collected.
- **How?** How is the value proposition created? Via the digitalization of analogue processes/machines using IoT devices; the collection, processing and storage of data collected by the IoT devices; and the provision of advanced analytics to predict and/or prescribe process or machine behaviour
- **Value?** How is value/revenue generated? For potential internal customers, examples could include improvement of operational effectiveness, reduction in operational costs, reduction of quality issues, improvement in reputation etc. For potential external customers, examples could include lower product costs, better quality, increased sales of existing products or services, and sales of IoT devices and services.

Business case phase

During the business case phase, the project team will conduct a thorough evaluation of the solutions generated, assessing their potential value to add impact and the resources required for implementation. The business case activities will begin with a session where each of the groups will present their top two/three solution concepts created in the ideation phase to the whole group. The aim of the presentations is for each team to pitch their solution concepts as a coherent and viable business opportunity that should be pursued by the organization.

Following the presentation of all solutions concepts, all participants should be given the opportunity to discuss the merits of each proposed solution before commencing the next activity. After the discussions are concluded, the workshop facilitators will co-ordinate an activity focused on the evaluation of each solution concept. This evaluation will consider two business-case aspects:

- The solution concept value-add potential (internal and external markets).
- The required resources for the implementation of the solution concepts and the alignment with the available resources (based on the outputs from the analysis phase).

The objective here is to make sure the solution concepts that have the best potential to add value to your business and that can utilize your existing capabilities and resources are identified as priorities in your digital transformation.

Everyone at the session gives the solution concepts a score based on these two criteria for evaluating the business case. This activity can be interactive and could use a larger four-box matrix with the two axes labelled 'value-add potential' and 'effort/resource alignment'. This allows standardized assessment and supports the team in reaching a consensus as to the selection of the solution concepts that demonstrated the highest potential value add and alignment with the business capabilities. A sample agenda for a business case workshop is given in Table 1, which you could use as a starting point for your own session.

Topic	Duration
Brief on importance of Industry 4.0	30 minutes
4 phases of practical Industry 4.0 implementation	15 minutes
Analysis phase results	15 minutes
Industry 4.0 internal capabilities summary	30 minutes
Ideation phase results	15 minutes
Solution concepts: presentation and discussions	60 – 120 minutes
Individual solution concept assessment (4 box chart)	30 minutes
Summary of business case phase evaluation	30 minutes

Table 1: sample agenda for business case workshop

Implementation phase

This is the final phase of your practical four-step approach to get your business started with Industry 4.0. In the implementation phase, the solution concepts with the best business cases are transformed into project proposals, presented to the senior leadership team for approval and, if approved, the solution is implemented.

What happens after the workshop?

The solution concepts and business cases evaluated in the workshop will need to be expanded into a project charter with a full set of specification to enable the implementation of the proposed solution. For this purpose, a project lead should be assigned and take on the responsibility for the further development and execution of the solution concept. First, they will co-ordinate a presentation for senior leaders to sign off. Then they will pull together technologies and resources with strategic partners and departments within the business.

From solution concept to value add

The solutions concepts and business cases will be presented by the respective project leads to your business's key stakeholders and decision-makers, hopefully these key people will already have been involved in the process from the beginning. If not, this is the time to engage and get buy-in. The proposed solutions could involve multiple areas across the business from design and engineering to production, sales and customer services. As such, it is critical that the key stakeholders commit to the Industry 4.0 activities proposed and see it as part of the business-wide strategy to profit from the implementation of Industry 4.0 technologies and solutions.

The introduction of practical Industry 4.0 solutions will allow you to extend your current business models in products, services and processes. This can be the first step in the journey to secure your position in your value chains and to develop new core competencies related to digital transformation and the creation of a comprehensive business strategy which includes Industry 4.0 projects and, as a result, you can benefit from future opportunities to improve profitability.

- *Templates, solutions and examples for managing each of the four stages of an Industry 4.0 project are available at: www.profitingfromi4.com.*

Notes

1. Available at: https://www.thegeniusworks.com/wp-content/uploads/2017/06/St-Gallen-Business-Model-Innovation-Paper.pdf

12.
IP IN CONNECTED VALUE CHAINS

As you advance in your Industry 4.0 implementation journey, you will move from digitalization projects confined to your business boundaries to projects involving the wider value chain. In such projects, the management of intellectual property in the connected value chains has the critical aim of providing a means of protection and controlled sharing of your intangible assets.

Unlike tangible assets, intangible assets such as data sets, knowledge, know-how, trade secrets and other form of valuable IP are difficult to physically control. Typically, once the intangible asset is out in the open it is almost impossible to control access and further distribution.

IP provides a set of mechanisms (patents, design rights, trade secrets, trade marks and copyrights) to protect intangible assets and manage who uses them. Thus, IP enables you to exert a level of control over intangible assets, making them the object of transactions that attract revenue to the business. This ability to rely on IP rights to provide control over intangible assets supports and enables their use and disclosure. It also creates an incentive to generate and invest in innovation, generating further IP and value to the organization.

Through the combined use of all appropriate IP protection mechanisms, you can leverage your innovation to recover the cost of innovation and secure a return on your investment. As such, IP should be seen as an effective means of securing long-term competitive advantage to support the commercialization of products, processes and services.

Industry 4.0 and the increase in networked innovation across the value chain poses a number of challenges to IP management. In theory, IP law provides an option to support the protection of intangible assets generated in a collaborative manner, such as the connected value chains, in the form of joint ownership and contractual agreements. However, there are a number of factors in Industry 4.0 networked innovation that complicate the use of traditional IP protection mechanisms. For example, the use of data sets, knowledge and other capabilities through connected value chains that naturally extend beyond your business boundaries can give rise to multiple businesses claiming to own the interests and rights in the same intangible asset.

Traditionally, businesses in manufacturing have relied on confidentiality and secrecy for the protection of intangibles where there is no clear legal protection mechanism available. This approach worked well in the past as most jurisdictions around the world have laws that protect confidentiality, and have sanctions on the disclosure and use of business secrets.

The protection of commercial information that is valuable and not in the public domain using confidentiality is broad, and the threshold for protection is a lot lower than that of a patent examination. Businesses with strong IP protection strategies normally use formal and informal protection mechanisms such as patents, trade secrets and contracts to achieve effective protection of their most valuable intangible assets.

The combination of formal and informal protection mechanisms is critical in the context of Industry 4.0 value chains where data exchange across the product or service lifecycle is ubiquitous. This creates an unprecedented level of knowledge codification and sharing, which if left unchecked has the potential to affect the appropriation of value from innovation created in your business, as well as your profits and long-term competitiveness.

To effectively address this level of openness between your business and the value chain, it is paramount that you identify what intangible assets are critical to your business value proposition and which are not. This will enable you to make informed decisions and formulate an effective protection strategy to dictate what, how, when and with whom, your key intangible assets are to be shared. This chapter will discuss:

- how to identify and protect intangible assets in Industry 4.0 value chains;
- how you can improve your awareness of potential risks;
- and how to take control over the sharing and use of your intangible assets.

I will begin with transactions in a value chain concerning existing intangible assets and will then discuss new IP generated collaboratively.

Existing or background IP

Intangible assets are a special class of property and, as such, they can generally be valued, purchased, licensed, transferred and sold. In addition to these traditional property transactions, intangible assets protected by IP law, such as patents and trade marks, can provide your business with the right of exclusion which enables you, as the owner of the intangible asset, to provide a licence to a third party interested in using the protected asset, normally in exchange for a benefit (licence fees or another commercial benefit).

Licences are normally used in collaborative projects by manufacturers when working in close co-operation with key customers and suppliers. As such, licensing should not be completely new to you, as it is the most common method for providing a means of guaranteed access to a particular intangible asset underlying an innovation or giving freedom to operate, eg, a promise by the IP owner not to sue a third party using a particular intangible asset or technology.

In a business-to-business innovation context, such as Industry 4.0 value chains, when you know the intangible assets required for a particular offering, your business may need to negotiate licences to guarantee access to technology. Likewise, you may be able to sell your IP or license it out to other businesses in the value chain who rely on your IP as part of their offering. This type of activity is commonly referred to as cross-licensing and can be used to secure freedom to operate and/or to avoid future risk of litigation.

Typically, licences are governed by law which may limit contractual freedom. For example, there are legal limits concerning activities, such as the use of disproportionate payments, the use of alternative technologies, reverse engineering or re-developing licensed IP. Furthermore, IP licence contracts also fall under the remit of competition law in most countries.

Nevertheless, as freedom of contract is embedded in most legal systems around the world, despite the above limits, licence agreements terms and formats can vary depending on the technology area, geographical area or the parties involved. As such, the ideal scenario is that each licensing contract should reflect the characteristics of the intangible asset, the IP protection mechanisms used and the commercial circumstances. However, in practical terms, model contracts and templates, including broad boilerplate clauses, are common practice across all industries.

Standard agreements

There is no doubt that as Industry 4.0 implementation progresses, these types of cross-licences and licence terms will become pervasive and will need to be standardized to include a larger number of businesses connected across the value chain. Business groups and industry associations are increasingly using these agreements to standardize the terms of contracts and licensing agreement.

There are a number of well-known examples of community standard licenses, such as the GNU General Public License (GPL) which is the foundation of software licences for open-source code. Also, the Creative Commons (CC) licence, which is commonly used across the internet in the context of copyright licences. These two examples of community-wide, mutually beneficial licensing standards are closely coupled with a code of conduct in which users or licensees agree to use the intangible assets in a particular manner which is aligned with the wider community's interests. This is clearly set out in the case of some open-source software license, where developers utilizing the open-source code are subject to a set of conditions which includes sharing any future developments and enhancements with the community. Such conditions on users of community licenses can also be found in recent initiatives in the automotive industry driven by companies such as Tesla, which famously promised not to sue other companies utilizing their patented technologies, provided that the companies themselves agree not to sue Tesla in the future for the use of their own technologies.

There are obvious downsides to these types of community licences, particularly the lack of control over future developments and the *carte blanche* you might give to a potential future competitor. The fact is that these types of licence are widely used. Many new businesses trading in complex ecosystems perceive community licences as a cheap and effective way to exchange and develop new technologies faster, which in turn accelerates their speed in introducing new solutions to markets.

In this manner, software solutions using community licenses are often developed in collaboration utilizing multiple sources, which may range from closed to partly open and fully open sources. This mix of sources and licensing terms and conditions behind commercial products and services often gives rise to complex challenges. Clarity about current and future business models is required before relying on equivalent community licences in the context of IP shared across Industry 4.0 value chains.

Limitations of IP for digital assets

The formal IP protection methods, such as patents, registered designs and trade marks, may look obsolete in the context of digital assets, such as product and process data sets or information technology. This can be attributed to the fact that technology is developing much faster than the IP laws around it. In turn, the degree of uncertainty increases the risks associated with the protection and the commercialization of new technologies.

This occurs due to the intrinsic complexity involved in digital assets, which makes it difficult to identify ownership of assets across a network of users and collaborators. In this scenario, businesses tend to become more protective of their assets and attempt to close themselves to collaboration. The licence agreements discussed above provide a means to address some of these shortfalls with traditional IP protection methods to promote safe collaboration between partners.

A value chain network for Industry 4.0 collaboration, for example, may benefit from a licensing agreement specifically for digital assets for which there are no other IP protection mechanisms. Such agreements could include a number of collaboration mechanisms which, in addition to the licence to use the assets, also covers aspects such as distribution of future costs to maintain critical assets and even the sharing of potential profits related to the future use of these assets.

Complex relationships across the value chain tend to evolve due to collaborators' changes of circumstance and often the agreements may no longer be perceived as a fair deal by everyone. For example, a collaborator may be acquired by a competitor, so you are faced with a relationship where shared digital assets may no longer be beneficial to you. It is important to have a contingency plan embedded into these agreements in the event of such situations. Industry 4.0 value chains will require increased flexibility and trust from all participants and a mechanism for agreeing on the adequate levels of contingency governing the use of intangible assets by the parties involved.

Collaborative codes of conduct and value chain rules should be considered and agreed upfront before your business begins the process of digitalizing your interactions with the value chain partners, particularly before the disclosure of any data or knowledge takes place. Your business should consider the use of a framework agreement to establish the main rules and limits of the collaboration to ensure any potential abuse by third parties as a result of data exchange in the value chain is prevented.

A framework agreement is the ideal mechanism to establish a clear and agreed set of objectives for the value chain and also to define an adequate code of conduct in relation to the use of intangible assets. It should include the use of any data, information and knowledge to avoid any misunderstandings from the start of the relationship. Unfortunately, opportunism and dishonesty should also be taken into account and the agreement must include clauses addressing potential issues related to mergers, acquisitions, takeovers and defection of collaborators.

In negotiating the agreement, an analysis should be made of how the clauses and contractual rules can be effectively implemented and monitored by all the partners. My suggestion is that you should consider best practice from other collaborative communities. These include the creation of a governing or steering committee to manage any disputes related to the process of collaborative use of IP and the use of emerging innovation. These committees are typically formed by representatives of all organizations involved, who will work to a defined administrative process for monitoring and reporting on any relevant changes, inputs and outputs related to the agreement.

This joint monitoring and reporting enables all parties to be fully informed of relevant changes to partners' positioning and potential changes of intent. As a result, businesses in the collaborative network can tailor their own positions to potential disclosures to prevent any detriment to their competitiveness in the value chain. Finally, the use of committees to govern complex collaborative projects is an effective approach to prevent and address effective co-ordination in case of potential conflict between collaborating parties. Chapter 15 will provide a number of suggestions on how to approach these agreements and the key questions you should consider in formulating them.

New or foreground IP

Agreements governing collaborative innovation initiatives, such as the Industry 4.0 value chains, require a clear set of rules regarding the use of existing IP at the beginning of the collaboration, IP developed during the relationship and, eventually, IP in the context of use after the conclusion of the collaboration or the exit of one of the collaborators. These aspects and rules should be agreed and established before the collaboration starts.

The agreement clauses usually deal with background IP, which refers to the use of intangible assets that you or your partners will bring into the collaboration at the outset. In essence, these would include any IP you own before you join the value chain. Background is broadly defined and can include all existing IP across all collaborating parties, including data, information, tacit and codified knowledge. It is important that your business identify and account for all of your existing IP as the set of background IP clauses will typically only apply to the assets included in the background IP register. You may also decide that there is some background IP which your business would like to keep outside the project and, as such, these should be intentionally excluded from the agreement's scope.

On the other hand, new IP generated during the collaborative initiative will also require a set of rules related to the use and ownership by the collaborators in value chain. These intangible assets generated after the commencement of the relationship are referred to as 'foreground IP' (an intangible asset directly related and resulting from the collaboration activities) and also 'sideground IP' (an intangible asset related to the topic of collaboration, but not part of the collaboration activities).

This latter aspect is an important consideration as your business will be engaged in other activities and potentially even other value chains where it will also generate IP outside of, and unrelated to, the collaborative relationships in this particular value chain. Thus, you should not only ensure that different types of foreground IP are covered as part of these agreements, but you should also be careful to ensure that there are no conflicts between different agreements in separate value chains which could result on restrictions on how you can use your foreground IP. Particular attention should be given to distinct levels of confidentiality and exclusivity applicable to each type of IP in the context of each phase of the collaborative lifecycle (before the collaboration agreement is signed, during the collaboration and after the collaboration is terminated).

Creating a complex agreement to govern the challenges and opportunities emerging from Industry 4.0 value chains and the resulting innovations is time consuming, but it will certainly pay off when it comes to appropriating value from digital transformation. The threat of sanctions and litigation for breach of contract will be your first line of defence against opportunistic behaviour from collaborators, but these measures are more effective when coupled with a detailed set of procedures and methods to execute the collaboration aspects pertaining to the sharing of IP across the collaborators. It is critical that all parties should understand how the rules in the agreement apply at a practical level in order to ensure adherence.

All parties should have a clear understanding of the value of the intangible assets and the related provisions to give adequate motivation for complying with the rules and support the ongoing collaboration between the parties. The combination of a threat of potential sanction with the promise of future opportunities is an effective way of ensuring a mutually beneficial relationship from a contractual point of view. However, this formal approach can and should be enhanced by building trust between collaborating partners at both strategic and operational levels.

The evidence I found as part of my research and practice in industry shows that businesses that understand well what IP they own and what IP is important to them have a better chance to proactively instigate the discussions and the rules of engagement for collaborative relationships, such as those in connected value chains. This improves their ability to secure the value emanating from existing and new IP.

If you are unsure what IP your business owns, how you manage it and what contractual rules apply, you are unlikely to benefit from long-term openness in Industry 4.0 value chains. As your business progresses with Industry 4.0, its experience with these type of complex collaborative relationships will increase. The knowledge about what IP is valuable to you and the preferred rules in such engagements will be consolidated as a core capability to ensure long-term success.

The remaining chapters in Part 3 will provide you with some insight and recommendations to help you leapfrog some of the hurdles that may emerge in this context and hopefully help you strengthen your position in terms of value capture from Industry 4.0.

13. IP STRATEGY FOR INDUSTRY 4.0

Why consider IP management when entering the world of digitally connected value chains? The activities associated with the creation, capture, management and use of intangible assets depends on many decisions that will have a lasting impact on what value and what profits you achieve with Industry 4.0.

The generation of such intangible assets is resource intensive in terms of time and cost, as well as in strengthening your reputation with partners and employees. At the same time, this type of asset can be fragile. You could quickly lose it in an inadequate collaboration project or by losing an employee with critical knowledge of the asset.

The complexity evident in open-innovation initiatives, such as collaborative projects in Industry 4.0 value chains, is not well represented in the literature and practice, as the traditional depiction of the innovation process takes the form of a neat funnel with clearly marked inputs, outputs and a set of symmetric relationships between the collaborators.

Industry 4.0 is based on integration and information sharing. In essence, it is an intensive exercise in open innovation. In this context, IP is not only about protection, but it is also about the controlled and strategic sharing of knowledge to advance your business strategy.

Traditionally, inventions were based on physical assets protected by patents and design rights. In the digital world, they rely as much on tacit and codified knowledge, such as data and digital information. The protection mechanisms in this case often take the form of secrecy, contractual agreements or carefully designed processes to secure information even within your business.

This chapter discusses the challenges for IP management in value chains that connect Industry 4.0 to collaborative open innovation. Greater interdependence in creating and transferring data and knowledge between partners in the value chain has implications for developing an IP strategy that maximizes the value that is created and captured from Industry 4.0.

IP strategies and collaborative innovation

There are many definitions of IP strategy and narrower cross-sections of IP, such as patent strategy. Nevertheless, the common feature between them all is that the purpose of an IP strategy is to guide the approach to utilizing a business's intangible assets to achieve its long-term objectives.

Academic definitions are not helpful in our context, as they mainly focus on closed innovation activities in scenarios where all the IP is generated as an output of your business's isolated efforts to innovate. As such, most definitions will focus on strategies designed to protect IP, excluding all others from use of the intangible assets in question. Little attention is paid to the strategies that require protection for the purposes of sharing IP in value chains with multiple partners, open sourcing and data exchange. As a result, the existing theories do not provide much support to successfully navigate the challenges and opportunities of digitally connected value chains.

In these new scenarios for value chains, we need a definition for IP strategy that accounts for both the protection and the sharing of intangible assets, seeking to achieve an optimum balance between the assets critical to your business's competitive advantage, whilst being open and inclusive to encourage interactions with other businesses in the value chain. This new definition for IP strategy should also include a clear hierarchical relationship between your business strategy, your business model and your IP strategy as a mechanism to help achieve the overall goals for your business. On this basis, I would like to propose a definition of IP strategy for your consideration in the context of Industry 4.0 as follows:

> IP strategy is your business's awareness of what intangible assets it owns, how important these assets are for your current and future business strategy, and how these assets must be protected, managed, traded and/or shared to enable your business to flourish.

IP strategies can, if appropriately crafted, link your business's assets, technologies and capabilities with your strategic objectives, making your intangible assets as visible as your physical assets. Furthermore, IP strategy also defines the protection mechanisms for each of these intangible assets, highlighting both formal (traditional IP rights) and informal (contracts, trade secrets and technical solutions, such as encryption and training) protection methods. It is important to keep in mind this wide scope when formulating IP strategy for connected value-chain innovation, as your business will be managing increasing amounts of digital assets governed by contractual agreements, policies, procedures and a number of other informal approaches.

Questions to ask

To put your business in the best possible position to capture value from digital innovation in the context of Industry 4.0, you should consider the following questions when formulating an IP strategy:

- What intangible assets are required to achieve your business objectives?
- How do agreements in the value chain impact the creation and use of these intangible assets?
- What are the pros and cons of connected value chains to your business?

- If you have external business partners, what are the options for creation and ownership? and what models for collaboration could you use?
- Finally, how can you draft and manage contractual agreements to align with different models of engagement with your partners in the value chain?

In my experience, traditional IP strategies in manufacturing fail to address these questions which are directly linked to strategies for maximizing how digitally connected value chains perform. For example, in the case studies discussed in Part 2, I found that the majority of businesses involved lacked even a general IP strategy. These quotes highlight common levels of underdevelopment across a variety of businesses:

- 'We don't have a specific strategy documents for managing IP … we have a kind of idea of what is important and what we should seek to protect from customers and partners.'
- 'No, we don't have an IP strategy. We have a central legal team who go through the key considerations on a case-by-case basis.'
- 'We are very flexible with our IP strategy, I am not even aware of a centralized IP list … I actually think we should pay more careful attention to this as we already have had a few issues with customers appropriating our IP and passing it on to our competitors at the end of a contract … IP is a big issue in our industry and will increase in complexity with digital supply chains.'

There are obviously businesses that are more mature and have a documented IP strategy that cover the process of identifying and disclosing physical inventions as part of the patent filling process. These documents, however, typically don't provide a link to the wider business strategy, so fail to support decision-making in the context of complex collaborative relationships.

I am not suggesting that the lack of IP strategy means that these businesses do not have valuable IP or that they are not committed to IP management in the long term. However, there was a strong correlation between the capability to innovate and capture its value with the presence of a robust and coherent IP strategy linked to the overall business strategy amongst the businesses interviewed.

Thus, my recommendation is that a systematic approach to IP management can be seen as a step in the right direction towards a stronger value-driven approach to manage decisions and improve your business's chances of profiting from Industry 4.0. Regardless of your approach to Industry 4.0 adoption across the value chain, it is paramount that your business focuses on IP strategy in the context of the connected value chains you are about to join.

Strategic options

There are many strategic approaches to IP management beyond the traditional sword and shield. Firstly, you can develop IP in different ways. Either you can conduct your own research and development to innovate. Or you can access IP through licensing and cross-licensing agreements with third parties to support your aspirations for growth and profitability.

You might choose to enter into a collaborative project to generate new IP. Or you could even collaborate with your customers who may be willing to contribute IP to your business, as they would ultimately benefit from your innovation.

There is also a wide range of things you can do with your IP once you have created and appropriately protected it. As a first traditional step, you could use your ownership rights to exclude others from using it. As an approach, it has its limits when innovating in connected value chains using digital assets. The most rewarding option for you in connected value chains is to find mutually beneficial licensing opportunities which will allow you to trade or even give away certain IP of limited use and keep the core IP safe through limitations built into the licensing agreement.

You could also make the IP available to your partners via an open access model in exchange for the right to use your partners' existing and future IP without the risk of litigation. This may help you access a wider market share or fulfil your business's vision and strategy.

All of these options could form part of your business IP strategy. Selecting the appropriate strategies to use in each scenario will depend on factors specific to your business. The best IP strategy will be flexible, taking into account different business models and the need to be open in entering certain collaborative relationships. Such an agile IP strategy will require you to consider different innovation stances that your business may take along the open to closed range, depending on a particular type of IP for a given type of relationship.

The most innovation savvy businesses often tailor different types of IP strategies depending on the situation. For example, the use of IP licensing strategies can help gain access to a new market, improve the quality of a current product or service, engage with future customers, and grow your business, industry or field as a whole. See for example the Tesla policy on open patents. It won't initiate patent actions against anyone acting in 'good faith', as long as they are not asserting any IP against Tesla or a third party.

This approach effectively means that Tesla's patents are only free to use if you do not enforce any right against Tesla or anyone else, you do not oppose Tesla's patents or copy its designs. As such, if you use Tesla's patents, you are essentially making your own IP redundant. On the other hand, Tesla benefits from the arrangement because it is free to use any improvements made to its technology by another party.

The key is to understand that IP can support you creating a sustainable and beneficial value-chain network where partners, including your customers and suppliers, can find mutual benefit. The most innovative and profitable Industry 4.0 examples may prove to be those in which customers and suppliers actively work together to innovate and deliver higher value to all stakeholders.

In the world's largest companies, the value of intangible assets is now widely recognized to be in excess of 85 percent of the total company value. Businesses are increasingly aware that licensing costs are high, but the costs of IP infringement can be even higher, as we have seem with the famous legal battles between Apple and Samsung.

On the other hand, if your business can become aware of new business models which enable it to use Industry 4.0 technologies not only to improve operational processes, but also to develop the IP in products, services and processes that you can use to generate attractive rates of return on your investment.

IP licensing is such a big business because it generates so much free cashflow. In manufacturing, it creates the potential for multiple businesses across the value chain to grow. In the next chapter we will discuss how to identify and control these intangible assets in a cohesive manner across your business.

14.
MANAGING INTANGIBLE ASSETS

In a revolution, such as Industry 4.0, the pace of technological and industrial change is not just rapid, it's unpredictable. To adapt and prosper, your business will rely on a number of techniques to identify and protect its intangible assets, especially those which are critical to your current and future value propositions.

Whilst there are a number of common strategies for your intellectual property, a one-size-fits-all solution does not exist, as your business will have a unique strategic objective and will conduct distinct activities to achieve its vision. As such, you can concurrently line up various asset types and protection mechanisms as part of a comprehensive IP strategy to support your Industry 4.0 endeavours.

As a result, your chances of profiting from Industry 4.0 innovation will improve, particularly if you engage your whole organization in the creation and implementation of your IP strategy, expanding it beyond your technical or legal departments. Typically, their focus is narrowly on protecting the business from the potential infringement of someone else's IP or from the infringement of your own IP by competitors. As discussed, these are not priorities going forward.

In response to the challenges posed by the integration of manufacturing value chains, your IP strategy will focus more on particular business models and the IP they require to achieve a particular value proposition in the context of your business's strategy. By doing so, you will be in a strong position to influence the formation and configuration of value chains by deciding in advance how to maintain a competitive advantage and retain your critical assets. At the same time, you will remain open and responsive to sharing data and collaborating in areas where you can innovate fast. This chapter sets out how you can formulate, adapt and align an IP strategy for Industry 4.0 at three distinct levels: strategy, policy and management.

Intangibles process design

Developing and executing a cohesive IP strategy is a difficult challenge, particularly when embarking into the uncharted territory of integrated value chains. Typically, businesses set their IP strategies at a high level. However, in order for these strategies to be effective, they must be cascaded and implemented at all levels within the business, across all functions involved in IP management decisions.

As discussed in the context of the case studies explored in Part 2, there is a clear disconnect between strategic vision and planning, and the actual business practices that lead to IP decision-making within manufacturing businesses. This leads to a lack of effective IP management in many businesses, where even the best efforts in developing an IP strategy can be ineffective if it is not built upon a foundation of policies, best practices and the appropriate management processes to execute the IP strategy on a practical level.

In each scenario, decision-making will benefit from support on all questions regarding IP in the context of your business's relationships. You will improve your chances of success if you can define the method of implementation and execution of IP strategy starting from high-level strategy all the way to the management processes. Your business's prospects for appropriating value in future will also increase if you consider carefully the method of execution for the IP strategy, as the policies and management activities require a more frequent level of review due to the nature of these new relationships.

Accordingly, it is recommended that you consider how to formulate and adapt your business IP strategies to address changes in the appropriability regimes of the manufacturing value chain at three distinct levels namely: IP strategy, IP policies, and IP management.

The IP strategy level

At the IP strategy level, your business will benefit from defining the high-level IP objectives in relation to specific target markets, business areas, businesses models, technologies and knowledge domains. These strategic objectives must be founded on your business strategy, which should be broken down into objectives for each specific business area offering guidance to employees at all levels.

Everyone will gain a coherent view of how your business generates and captures value from each business area, as well as how the IP strategy supports the specific business model for each business area. Similarly, IP strategies will be tailored to business areas that may be in different geographical locations and subject to distinct jurisdictions. Entirely different business models may also be covered, such as manufacturing consultancy, engineering services, product manufacturing or other related services, such as aftermarket parts production and distribution.

Furthermore, you will identify the strategic objectives that require particular technologies, ensuring you align your technology roadmap with your IP strategy. You can then support the protection of technology and knowledge to achieve your desired business models for each area, not just for the current business, but also for the future.

Alignment to the technology roadmap

The importance of a robust technology roadmap supported by IP surveys of the technological areas should not be underestimated when considering new technologies and the development of new knowledge. Such a roadmap should be composed of a clear development and acquisition plan for onboarding these technologies.

Each business area will have its own requirements for technology and knowledge. Only a proportion of it will be subject to formal protection with registered IP mechanisms. For example, the protection of data sets or of inventions based on artificial intelligence are far from straightforward.

The same is true for different types of competencies, where the formal and registered IP mechanisms for protection do not amount to an effective and desirable solution. These include cases where there are high costs associated with the formal protection or where there are problems related to areas of technology prone to fast-paced development or in cases where the best option for protection is to keep the knowledge confidential and rely on trade secrets

The IP policy level

In order to align your business strategy and the IP strategy, you will have to ensure that suitable IP policies are created to define the appropriate protection mechanisms required to achieve the IP strategy's objectives and support the business model of a particular business area. These policies and mechanisms are likely to vary between different business areas and business models. Care has to be taken in their selection, aligning them with the business model and the appropriability regime.

IP valuation

At the policies level, you will be looking to value your IP, firstly as an asset and then in respect of how it aligns with your strategic aims. IP assets are difficult to evaluate without the business model which aggregates and captures value from the particular intangible asset.

This difficulty in evaluating IP assets without the business model context arises because there are many ways in which you can utilize IP for value appropriation. For example, IP can guarantee a monopoly for a product in a certain market in the case of patents, but, at the other end of the scale, IP has embodied know-how protected by trade secrets that can be used as an asset in collaborations with partners lacking that particular knowledge.

The valuation of IP is also difficult because two actors in the value chain will often have different views regarding what it is worth as an asset. This is mostly because different businesses will have distinct business models and therefore different views of the business potential that could be realized by utilizing the IP assets in current and future commercial deals.

More importantly, IP assets can be used as proxy to an insurance policy and enable your business to enter strategic markets without the risk of infringing other business's IP. In these examples, IP is valuable in the context of the business strategy in integrated value chains.

Finally, also in the context of integrated value chains, IP can act as a badge of quality and innovativeness which can attract potential partners and collaborators who seek to develop products and services in collaboration. The IP developed in connection with customers' and suppliers' products or services which are commercialized locally or globally can be a valuable source of additional revenue for your business, if you can benefit from licensing IP to customers, suppliers and other third parties.

The IP management level

The IP management level refers to the day-to-day processes and activities which inform your business's decisions with regard to IP creation, evaluation, protection and commercialization. The effectiveness of these processes will result in the success of the overall IP strategy. Likewise, any failure at the process level should inform changes to the IP management policies and processes to ensure that the IP strategy objectives are achieved.

In the dynamic environment of integrated value chains where IP in the form of data sets, drawings, 3D models, operational information and many other intangible assets are exchanged across businesses in the value chain, you will benefit from ensuring that your team has the appropriate IP management processes, which include the assessment of IP assets in the context of the data shared, as well as the data received by other businesses in your value chain.

Such an assessment is important in order to understand changes to the context in which the IP asset is being commercialized, as these changes might impact significantly upon the internal and external conditions, which can also influence the IP strategy. This assessment will ensure that the particular management approach and that the IP in question is still relevant to the IP and business strategies.

Aligning business, technology and IP strategies

Your capture of value from Industry 4.0, particularly as you enter integrated value chains, will improve if you create a coherent link between the business, technology and IP strategies, as it will support the selection of the most effective portfolio of protection methods to secure the intangible assets critical to achieving your strategic objectives.

The IP strategy should explore the technologies identified in the business strategy, defining which technologies and knowledge are core to the manufacturer's value proposition and how these IP assets should be created, evaluated, protected and commercialized. The IP strategy should also identify which technologies and knowledge are not core to the manufacturer's value proposition and which in turn can be shared and/or disposed of.

The range of protection mechanisms required in the integrated value chains will include formal and informal methods. Special attention should be given to contractual agreements and practical means of protection, such as encryption and access restrictions to critical information.

In the past, knowledge protection strategies across different businesses were often similar. However, the actual IP strategies for protecting knowledge in digitalized manufacturing value chains must be tailored to your particular business strategies and models.

Patent-based IP strategies provide strong protection. However, this type of strategy is only effective for certain technology areas with certain lifecycle characteristics. For example, in the pharmaceuticals industry, due to the long product cycles, patent-based strategies are an effective method of protection and can provide the patent holder with a monopoly that in turn leads to a competitive advantage.

In comparison, in areas of technology with a shorter product lifecycle, the same strategy is no longer effective as the period between a patent application and grant (typically two to four years) means that the technology could become obsolete before the monopoly is granted.

The formal protection mechanisms for IP protection are still an important part of the IP strategy toolbox for manufacturers. For example, patents will be important for manufacturers where they wish to gain exposure and improve the value of their brands by being recognized as a leader in innovation in a particular field.

However, informal protection mechanisms can be a primary source of value and benefit in the context of data exchange in integrated value chains. Methods such as publishing innovations or faster innovation cycles bring significant protection and advantage for certain technological areas.

For example, publishing or sharing a particular innovation or technology know-how is a strategy used when they are core to your value proposition, but could attract collaborations that lead to the development of products or services. As a strategy, you are creating a pull for collaborative research and removing the barrier of novelty to patents by your competitors.

Finally, secrecy is an important weapon in the protection arsenal. This mechanism relies on identifying and maintaining technologies and knowledge to avoid sharing both internally (amongst employees and contractors) and externally (with customers and suppliers). Once again, the emphasis is on identifying the critical knowledge required to achieve the desired value proposition and controlling such knowledge tightly. It is a key function of the IP strategy to identify who should have access to what knowledge internally and externally.

Contractual protection

Contracts play a lead role in protecting your IP in integrated value chains. Regardless of your particular businesses model, you are likely to find yourself increasing data exchanges and collaborative activities with your partners in the value chain, all of which are formalized through contractual agreements.

Careful consideration of potentially onerous clauses in old contracts which render IP appropriation impossible will allow you to improve the chances of future value capture and your freedom to operate. Such contractual clauses are common practice in manufacturing and were evidenced in the data collected in the case studies discussed in Part 2. In these scenarios, the manufacturer agrees to transfer any IP generated in relation to a particular product, service or process to the customer, including all formal and informal IP.

Unless your business model does not rely on value generation from IP assets, you should avoid such agreements and ensure that they are re-negotiated before embarking on the creation of integrated value chains. Failure to do so will mean that your business will be placed in an extremely weak appropriability regime where you will be unlikely to appropriate any value from Industry 4.0 innovation beyond the short-term benefits gained by operation improvement.

On the other hand, you will also be in a better position if you can include clauses on confidentiality, ownership of background and foreground IP, definitions of rights to use all forms of IP, non-competition clauses and other forms of protection as part of your business's future agreements.

Such agreements between your business and its partners only protect IP externally in relation to contracting parties. However, you should also be aware of IP protection in relation to members of staff, as the risk of knowledge leaking is greater due to the codification and sharing of knowledge across departments. This form of internal IP protection can be governed by strong contractual clauses in employment contracts to prevent employees from breaching confidentiality, appropriating inventions and competing with the manufacturer.

In the next chapter, we will explore a set of recommendations for addressing the practical challenges of profiting from Industry 4.0 with a view to improving how you appropriate value and develop your IP strategies in integrated value chains.

15.
IP ACTIONS FOR DIGITAL MANUFACTURERS

Depending on the capability of your business as an innovator, you might think that intellectual property only matters for large businesses in industries such as biotechnology, computer software or publishing. IP certainly matters in these areas, but, as emphasized throughout this book, Industry 4.0 is opening and connecting value chains, improving productivity and accelerating the knowledge economy, as well as the significance of IP strategy for manufacturers.

IP is context dependent, so it means distinct things to different businesses. Although it generally takes the form of rights that vest in creators of innovative ideas, expressions, processes and brands. Every business will have a different IP portfolio of some kind and value, whether or not they recognize it. As a consequence, every business requires an IP strategy, even if it is a simple one.

This is easy to appreciate if you consider the example of IP contained in the name or brand of a business. Every business has a brand that describes it to potential customers and partners. No matter how small or large you are and in which industry you operate, the brand represents a significant form of IP. Equally, most businesses are learning to operate in an IP environment in which industry revolves around the production and management of information in the form of intangible assets rather than physical production and physical assets.

The more information your business has to create and manage, the more likely it will be operating in the realm of intangible assets, as opposed to physical assets.

In the same way as the information economy is exploding in size, IP is rapidly gaining prominence. It is being tested and is developing constantly at the intersection between traditional protection mechanisms and new technologies. Just look at how artificial intelligence is affecting patent law in the case of AI-generated inventions or copyright in the case of AI-generated artwork. These are the reasons why a well-considered and flexible IP strategy is paramount to giving your business an edge when combined with Industry 4.0 implementation.

IP itself is changing in terms of its everyday practice and its social norms, as well as the way in which businesses think about it and act on it. Its marketplace is becoming ever-more global with expanding opportunities to capture value. Large countries like India and China are fast moving from providers of cheap labour to becoming lead IP creators. In this global shift towards a knowledge economy, IP is increasingly recognized as the currency for the ecosystems it creates.

In this environment, the management of IP is more complex than other asset classes. Depending on your business strategy and your portfolio of products and services, IP might seem more like a minefield. Its multiple facets in the context of future value chains emerging from Industry 4.0 requires a careful and proactive approach. Regardless of your size, industry or maturity, you now need to think about IP more than your predecessors once did.

The technologies are changing, your value chain is changing and the law is changing, sometimes rapidly and on a global scale. The biggest changes are coming in how business leaders are thinking about IP in strategic terms. Strategy in this area is dramatically different today than it was even ten years ago.

Your future prosperity depends on establishing an IP strategy capable of being resilient and flexible enough to cope with the current and future challenges of the global knowledge economy and the irreversible transformation brought by Industry 4.0.

Formal and informal exchanges of IP

You should carefully consider formal and informal exchanges of intangible assets in the connected value chain and ensure your business is aware of the consequences. Formal exchanges of intangible assets normally take place in the context of formalized and legally protected IP, such as buying, selling, and licensing patents. These formal IP exchanges are possible when all the relevant assets and the partners can be identified.

Open innovation or collaboration across the value chain is likely to be less transactional, allowing for an increased flow of data and more informal exchanges in the sense that intangible assets are not always identified. These informal exchanges promote the sharing of non-traditional forms of IP and work well when there are good rules concerning the collaborative relationship between the parties, for example, rules governing the aim of collaboration, and the process for background and foreground IP protection and ownership.

The ground rules concerning collaboration in value chains may be created by a framework agreement. However, you should remember that these informal exchange may not be always beneficial to your business. Uncertainty about the assets to be shared (background IP) and the assets created in the process (foreground IP) could subject the entire collaborative network to opportunism and free-riding.

The informality of the exchange makes it harder to evaluate the impact of potential IP appropriation by other businesses in the value chain, as well as easier for a collaborator to exit. Clear contingencies and provisions relating to voluntary or involuntary exit should be set at the beginning of the relationship and actively monitored.

Your business will benefit from developing an in-depth understanding of collaborative contracts. Significant efforts are required to optimize collaboration frameworks and put supporting processes in place.

Despite the paramount importance of contracts, they are still neglected by many businesses, especially at the beginning of the process of collaborative innovation. The reason for ignoring contracts is that they are normally complex documents which require the early consideration of essential information by the collaborating parties, such as a list of background IP. As we have seen, some manufacturers do not have such a list of intangible assets. Another key perception is that contracts are formal legal devices which add bureaucracy to relationships and prevent the free flow of innovation. However, as an interviewee pointed out in one of the case studies:

> Typically, the collaborative relationship goes wrong when partners are not upfront with their background IP and you can't reach an agreement on the terms. The negotiations can be very painful and delay the project initiation sometimes. However, I don't think any innovation project has ever failed because of the collaboration agreement. In fact, it helped us to identify the IP we had and to argue our rights in the foreground IP.

From an empirical point of view, I would argue that you should look at contracts as a method to strengthen your traditional (formal) IP protection and also as an alternative to formal protection where it's not available, let's say in the case of intangible assets in the form of data sets of AI algorithms which are unlikely to be protected by formal IP mechanisms such as patents.

The dynamic nature of connected value chains will make it difficult for your business to rely on traditional IP protection alone. In the absence of a legal safeguard for your background IP, these value-chain relationships will require you to be strategically involved in the governance of IP and knowledge within the network, closely monitoring changes to the appropriability regime. Contractual agreements give you a powerful tool to govern these future relationships.

Use informal IP protection methods

Non-disclosure agreements are one of the primary contractual devices for supporting intangible asset exchanges and allowing a dynamic flow of information between collaborators. As a Tier 1 manufacturer commented in one of our interviews:

> NDAs are the most common method we use to protect our IP. Every time we talk to a customer or a potential partner, we use them. This is very helpful to create a level of confidence and get the co-operation going whilst we get to know each other's strategies and commercial angles. We use standard forms and it is part of our process.

Even though NDAs are common, not all manufacturers use them in the context of data or information exchange with third parties. In addition, the scope of NDAs should also be tailored to a particular scenario. They can extend from a simple non-disclosure of information received all the way to a non-compete clause in your particular markets.

It is best to make NDAs proportional and reasonable considering the opportunities and benefits of the collaboration. If the terms are unreasonable, your business should re-negotiate and push for a more balanced position, as highlighted by an engineering consultancy in our interviews:

> Sometimes the NDAs terms impose that we cannot work for other manufacturers in this sector during a long period, even for 24 months once. We simply cannot sign them, as if we did, we would not have a consultancy business. In one case we had an NDA in relation to a consultancy project for ten working days and the customer wanted to block us from working with other companies offering the same product for five years. I don't think this is reasonable, but there are businesses, especially those at the top of the value chain who will always start with this negotiating position.

On the other hand, you should also consider that if open-source software is utilized there is a mandatory condition as part of the licence which means the derivative codes will have to be made public due to the GPLs (general public licences). This could cause an issue if your business plans to offer a solution based on open-source software to a customer demanding a confidentiality agreement or NDA. This is also the case in relationships involving academic and research organizations who may have a requirement to publish the findings of their research, which can result in friction between partners.

In collaborative settings involving industrial partners, there is an added element of pressure on academics and researchers working on innovation initiatives which may contribute to human error in the form of non-adherence to IP processes designed to safeguard intangible assets. An example of this in when an employee mistakenly shares confidential technical information with collaborators.

This renders statutory or contractual remedies that are typically effective threats to intentional IP infringement quite ineffective. The utilization of confidentiality agreements and NDAs should be part of a wider strategy to educate, encourage, monitor and support your team to understand how to appropriately protect trade secrets.

You should consider that even in closed innovation models within a single business, there are risks related to common practices, such as the use of external expert resources and collaborators in the inventive process, leading to the emergence of issues during IP protection or commercialization. Businesses that have realized the importance of open innovation models often make a strategic commitment to invest in developing capabilities and framework agreements to enable collaborative innovation with their partners.

In certain cases, business use alternative defence strategies, such as proactively publishing a particular technology to guarantee their freedom to operate and prevent other businesses from filing patents. To strengthen your position, this method can be used in combination with other informal means of protection, such as recruitment freezes, non-compete clauses, access limited to those who need to know or speeding up the innovation cycle.

Not all valuable knowledge can be isolated and protected as IP. Some of the knowledge in your business will collectively reside with individual employees. This includes the collective tacit knowledge or aggregated knowledge distributed across the business in your specialist teams.

Using IP law to obtain the rights over a particular intangible asset is one way of isolating and protecting valuable intangible assets. However, collective and individual knowledge is difficult to isolate and even harder to protect using traditional IP strategies such as patenting. Another way to achieve protection is through physically and/or contractually controlling and isolating a particular intangible asset. For example, a regular process to capture and transfer knowledge from employees to the business, contractually binding them to non-disclosure, is a highly effective way to secure IP.

As well as IP policies, other means to control and protect knowledge sharing may have to be considered. A confidentiality clause, which is often aligned with internal policies and through non-disclosure agreements with the collaborating parties, is often used to complement IP policies.

Value from data

The success of Industry 4.0 depends on vast amounts of data shared and aggregated across the entire manufacturing value chain. The rights to the individual data sets, as well as the bigger aggregated data sets, and the knowledge and information emanating from them will be critical to how manufacturers perform.

In order to address the challenges related to the paradox between the need to share with the need to protect your valuable intangible assets, a non-exhaustive and general set of actions that can improve the appropriability position and the protection of your businesses' intangible assets in the integrated value chain and in relationships involving data exchange is provided below.

- Establish IP as a valuable asset class (rather than solely as a sword and a shield), understand its value in the short and long term, then manage it, finding ways to expand the range of opportunities to achieve your core mission, whether that is profitability or another outcome.

- Be open to what your customers, competitors and others can offer you in terms of IP. Your most important IP might come from unlikely sources. In many instances, your business will be positioned in a value chain that can support the growth and profitability of multiple participants if the ecosystem itself thrives in a collaborative environment.
- Start from the premise that IP is most valuable insofar as it creates freedom of action and improve the collaborative attractiveness of your business, rather than serving as an offensive weapon against others. As a related concept, you should understand the extent to which your brand value is connected with your IP portfolio, your innovation capability and your reputation.
- Establish a strategy that lets you be creative and flexible in what you do with your IP by thinking beyond defence-and-attack strategies, giving you the scope to harness the power of increased openness and interoperability.

Different types of data are shared in the inter-organizational relationships that result from Industry 4.0 and integrated value chains. Each requires protecting. In contracts between manufacturers and the value chains, the following exchanges should be covered:

- **Raw data, machine data and unprocessed data**: this type of data refers to research and development and also to operational data generated by machines and other devices.
- **Analysed or processed data**: this type of data refers to the analysed or processed data generated or held by any business in the value chain (suppliers, manufacturers, customers, end users etc).
- **Manual or input data**: this type of data refers to the manual data input by users of connected machines and business systems across the value chain.

Data ownership and protection

Your business should also consider the fact that, similar to joint IP ownership clauses, data ownership and rights clauses contained in contracts governing the value-chain relationships will be the subject of much negotiation. These will often be contentious, as the powers of the various parties in a value chain will influence how much each party will be commercially pressured to share or transfer. Nevertheless, such contracts and their clauses governing the sharing and ownership of data should at least consider the following ownership, rights and licensing constructs:

- What data is subject to the contract?
- What rights are allocated to which party to the contract?
- What specific IP is owned or licensed to which party?
- Who is the licensor and who are the licensees?
- What are the licensees' particular business?
- What products do they offer and in which industries do they operate?
- In what territory?
- What is the term (time) of such a right?
- Are the rights exclusive or non-exclusive?
- Is there a right to sub-license?

Finally, you can also benefit from defining the expectations, responsibilities and liabilities regarding data security and privacy, as both suppliers and customers could increase the chances of a cyberattack resulting in a data breach. The contracts should incorporate such expectations, responsibilities and liabilities clearly and carefully. They should include the details regarding gathering, anonymizing, notifying and using suppliers', partners' and customers' data.

A careful evaluation of your particular business strategies and the impact on your business's appropriability position in order to select and include these constructs in the particular contractual agreements with suppliers, partners and customers will improve considerably your potential from value capture form Industry 4.0 innovation.

In connected value chains, when you neglect, ignore or fail to consider IP as an important asset, you are likely to be positioning your business at a disadvantage and creating unnecessary risks. You will certainly be missing opportunities to grow and damage your competitive advantage as manufacturing industry fully embraces the age of the knowledge economy.

By considering the actions outlined above, you can begin to mitigate these risks. Whilst doing so, you are likely to lead your business in the right direction and improve your innovation capabilities, whether that is manifested as higher profits, a stronger balance sheet, better products, or closer relationships with your customers and business partners.

16.
A MODEL TO PROFIT FROM INDUSTRY 4.0

Industry 4.0 has many potential advantages to you and your business. However, with these opportunities also comes the risk of an adverse impact on appropriation regimes, particularly for manufacturers down the value chain. As we saw in the case studies in Part 2, their intangible assets are being affected to varying degrees, depending on their area of technology or knowledge.

Within the integrated value chains that Industry 4.0 encourages, intellectual property is becoming a primary means for appropriating value and securing a return on your investment in innovation. As such, a tailored IP strategy is a cornerstone of your Industry 4.0 implementation. You can design it to account for differences between business areas and business models, generating a portfolio of IP assets, each with the necessary protection mechanisms and strategies for commercialization or sharing.

Such coherence and agility will put you in a better position to appropriate value from Industry 4.0. Your aim will be to match the technological competencies required to remain competitive in your current target markets, as well as anticipate what future demands could be.

With this in mind, we conclude with a development framework for Industry 4.0 and your IP strategy, so you are ready to address the main issues for operating profitably in integrated manufacturing value chains. This framework also brings additional benefits in aligning and communicating business and IP strategies across the organization. The framework links business strategies to business models and to IP strategies as depicted in Figure 1.

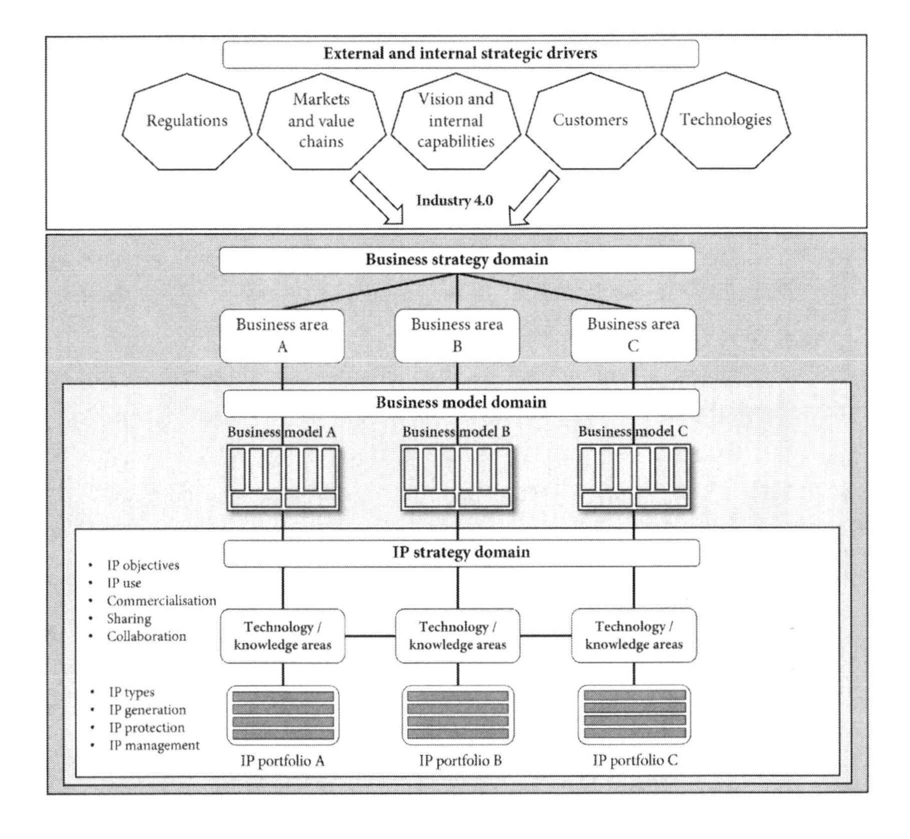

Figure 1: business and IP strategy integration framework

The framework shows the journey which begins at the top of Figure 1 with an evaluation of the external and internal drivers influencing your business strategy. Once these factors are taken into account, you can create an overarching business strategy which is cascaded into each business area which will execute the business strategy via a particular business model.

From the business model, shown in the business model domain, you will be able to identify the core technological or knowledge areas required to achieve a particular value proposition. Once these areas are defined, you may want to establish the necessary level of control over the required IP (the IP strategy domain) to enable the successful functioning of the particular business model.

Furthermore, your business may also decide to tailor the level of resourcing and investment in these areas. You can also identify where certain IP assets are not required by the business so they can be sold or licensed.

Your business can become clearer and more precise regarding its appropriability stance in partnerships, relationships and collaborations with different IP assets relating to distinct technology and knowledge areas. Collaborative developments will rely on such clarity.

In certain circumstances, due to existing knowledge, capabilities and strategic positioning, you may choose to avoid certain relationships in the value chain which may lead to conflicts and loss of competitive advantage. This typically occur when partners are also competitors with similar knowledge, capabilities and strategic positioning in the value chain.

Your business will be in an advantageous position by proactively identifying and classifying intangible assets in order of importance, both the technologies and knowledge you currently require, as well as those for the future. This identification is based on current and future markets you either serve directly as a business or indirectly via IP licensing.

Technologies and competencies may also be required for future business models. For example, the case studies in Part 2 required IP in technology and know-how across materials, products, processes and information technologies.

On the other hand, a single IP asset can have intrinsic value for multiple business models within the same organization and your business can identify how each individual IP asset relates to each business model within the overall business strategy. IP can then be protected and managed more effectively.

In addition, a coherent IP strategy will define which IP assets to acquire from external sources, which to commercialize by transferring to external partners and which to develop through collaborations. This comment from a Tier 1 manufacturing manager gives an example of the benefits:

> We had worked on this collaboration project with a customer and shared a lot of knowledge which enabled our team to identify an improvement to the product which was co-developed with the customer.

Such positive outcomes depend on a wider consideration of the entire business case for the network and each partner's relative position. Stronger protection of IP by an individual manufacturer will not necessarily lead to a benefit to the wider value chain.

The Marc Model developed as part of the research used as the basis for this book is designed to support the formulation of your future IP strategy by providing a method to perform an assessment of the appropriability regime and to consider, given particular business models and IP strategies, the likely position of each business in the immediate value chain and, as a result, the likelihood of value appropriation from innovation in a particular scenario.

The model can also be used in your business to understand the effectiveness of both formal and informal IP protection mechanisms and strategies that will depend on what business strategy and business model you adopt. As such, patents may be effective for one manufacturer given its wider context (business strategy, business model and particular areas of technology), but not at all effective for other manufacturers to whom patents may only bring unnecessary costs and future risks related to patent litigation.

This will help you make decisions on how to act in certain business models where informal protection mechanisms and the use of well-crafted contracts is the most effective method of protection and should be a more prominent part of a manufacturer's IP strategy as it can offer better protection to digitalized knowledge contained in data sets and know-how.

The proliferation of technologies improving operational efficiencies and the harmonization of systems across the value chain is intensifying competition. So manufacturers may increasingly adopt strategies to commercialize complementary IP via licensing agreements.

They can also open up alternative source of revenue by licensing non-critical innovations within their value chains. In such a strategy, formal protection mechanisms, such as patents, are still relevant and should support manufacturers in appropriating value in horizontally integrated value chains. This is a common thread across many high-value industries, where increasingly manufacturers and their collaborators successfully develop and commercialize technologies via licensing to other manufacturers who are interested in making and commercializing the physical products in the same or in an alternative value chain, market or industry.

Closing remarks

It is undisputed that IP is becoming increasingly important for manufacturers in collaborative settings. It is leading to an alignment of manufacturing with other high-value industries where technology development relies on complex collaboration and cross-licensing of IP rights. However, another challenge to manufacturers is that, particularly in the manufacturing value chain, the practices surrounding IP use and commercialization are somewhat immature and IP ownership can be seen as a drawback.

Successfully delivering the recommendations required to transform your business will not be easy, but it certainly will pay dividends in the future by building a strong foundation to support your business's mission. Recognizing the need for change and knowing when to take action is key. Often the symptoms show themselves long after the problem has taken hold: by the time an organization realizes a particular area or process is in trouble or, in this case, when your

Industry 4.0 investment gives signs that there is little value to the business, the problems are already deeper than you think.

When this happens, improvement and transformation have to be made a priority. Actually, figuring out what you need to do and where you need to be, can be one of the most difficult elements of transformation because it requires genuinely creative operational and commercial thinking, which can be a scarce resource.

Having a team prepared to take responsibility for implementing transformation is one thing, but never underestimate the importance of having a leader who knows what qualities are required, which ones they have and how to fill any gaps there might be.

The complexities of Industry 4.0 implementation across the value chain could mean that a change in one organization may have an impact on other businesses across the network. To succeed, your Industry 4.0 journey will have to be truly integrated and look at all parts as a whole, especially people and the commercial dynamics present in each area of the value chain.

Finally, change is normally slow and organic. Transformation, however, often involves getting five years of change done in six to twelve months, leapfrogging many of the organic adjustments you might otherwise make. It is important to build this expectation of pace into your expectations. Hopefully, this book will contribute to setting you and your business on the right path to profit from Industry 4.0.

GLOSSARY AND ABBREVIATIOINS

Application programming interface (API): an application programming interface is a connection between computers or between computer programs. It is a type of software interface, offering a service to other pieces of software. A document or standard that describes how to build such a connection or interface is called an API specification.

Appropriability regime (AR): the term appropriability in this context is used to characterize the extent to which the manufacturer innovating is able to obtain a return equal to the value created by that innovation.

Business models (BM): a representation of a firm's underlying core logic and strategic choices for creating and capturing value.

Closed innovation (CI): refers to an innovation system that takes place exclusively within the business.

Creative commons (CC): creative commons licences give everyone from individual creators to large institutions a standardized way to grant the public permission to use their creative work under copyright law. From the re-user's perspective, the presence of a creative commons licence on a copyrighted work answers the question, 'what can I do with this work?'.

Cyber-physical systems (CPS): term used to refer to integrations of computation, networking and physical processes. Embedded computers and networks monitor and control the physical processes with feedback loops where physical processes affect computations and vice versa.

Digital twin (DT): a digital representation of a physical asset, sufficient to meet the requirements of a set of use cases for which it is created.

General public licences (GPL): a general public licence is a series of widely used free software licences that guarantee end users the freedom to run, study, share, and modify the software.

Horizontal integration (HI): term used to refer to a network of vertically integrated and optimized businesses enabling data exchange and visibility for efficient value creation across the value chain.

Industrial internet of things (IIoT): refers to the extension and use of IoT in industrial sectors and applications. With a strong focus on machine-to-machine (M2M) communication, big data and machine learning.

Industry 4.0 (I4.0): term used to refer to the Fourth Industrial Revolution.

Information technology (IT): refers to the study or use of systems (especially computers and telecommunications) for storing, retrieving, and sending information.

Intangible assets (IA): refers to an asset that is not physical in nature. Goodwill, know-how, knowledge, brand recognition and intellectual property, such as patents, trade marks, and copyrights, are all intangible assets.

Intellectual property (IP): refers to creations of the mind, such as inventions, literary and artistic works, designs, and symbols, names and images used in commerce.

Internet of things (IoT): the connection via the internet of computing devices embedded in objects, enabling them to send and receive data.

Manufacturing appropriability regime construct (Marc): refers to a bespoke method used in this context to understand the business's ability to capture the value generated by Industry 4.0 innovation in the connected value chains for manufacturing industry.

Manufacturing readiness levels (MRL): manufacturing readiness levels is a method for estimating the maturity of manufacturing process and technologies, similar to how technology readiness levels (TRL) are used for technology readiness.

Non-disclosure agreement (NDA): a non-disclosure agreement is a legally binding contract that establishes a confidential relationship. The party or parties signing the agreement agree that sensitive information that they may obtain will not be made available to any others. An NDA may also be referred to as a confidentiality agreement.

Open innovation (OI): refers to an innovation system based outside the business boundaries in order to increase one's own innovation potential. Own employees, customers, suppliers, LEAD users, universities, competitors or companies of other industries can be integrated in the process.

Operational technology (OT): operational technology or OT is a category of computing and communication systems to manage, monitor and control industrial operations with a focus on the physical devices and processes they use.

Original equipment manufacturer (OEM): refers to a company that manufactures and sells products or parts of a product that their buyer, another company, sells to its own customers while putting the products under its own branding.

Product lifecycle management (PLM): refers to the processes related to a product as it moves through the typical stages of its product life: eg, design, manufacture, supply, service and recycle/reuse of a product.

Research organization (RO): refers to an entity, such as university or research institute, whose primary goal is to conduct fundamental research, industrial research or experimental development and to disseminate their results by way of teaching, publication or technology transfer.

Return on investment (RoI): refers to a performance measure used to evaluate the efficiency or profitability of an investment or compare the efficiency of a number of different investments. To calculate RoI, the benefit (or return) of an investment is divided by the cost of the investment. The result is expressed as a percentage or a ratio.

Robotic process automation (RPA): robotic process automation is a form of business process automation technology based on metaphorical software robots or on artificial intelligence /digital workers. It is sometimes referred to as software robotics.

Small and medium-sized enterprise (SME): the UK definition of SME is generally a small or medium-sized enterprise with fewer than 250 employees. While the SME meaning defined by the EU is also businesses with fewer than 250 employees, and a turnover of less than €50 million or a balance sheet of less than €43 million.

Tangible assets (TA): refers to an asset that has a finite monetary value and usually a physical form.

Technology provider (TO): refers to a company providing the technology and associated knowledge required to develop, test and/or manufacture a product.

Technology readiness levels (TRL): technology readiness levels is a method for estimating the maturity of technologies during the acquisition phase of a programme, developed at NASA during the 1970s. The use of TRLs enables consistent, uniform discussions of technical maturity across different types of technology.

Tier 1 supplier (Tier 1): refers to a company that manufactures and sells products or parts of a product directly to an original equipment manufacturer.

Tier 2 supplier (Tier 2): refers to a company that manufactures and sells products or parts of a product to the Tier 1 supplier.

Value chain (VC): a set of activities performed by various companies operating in a particular industry in order to create and deliver value to its customers.

Vertical integration (VI): term used to refer to the integration of business processes vertically within a single organization (sales, engineering, production, logistics and finance, among others).